Springer Series in Solid-State Sciences

Editors: M. Cardona   P. Fulde   K. von Klitzing   H.-J. Queisser

G. Venkataraman　D. Sahoo
V. Balakrishnan

Beyond
the Crystalline State

An Emerging Perspective

With 87 Figures

Springer-Verlag Berlin Heidelberg New York
London Paris Tokyo

Dr. Ganesan Venkataraman

ANURAG, Defence Research & Development Organisation, RCI PO
Hyderabad 500 269, India

Dr. Debendranath Sahoo

Indira Gandhi Centre for Atomic Research
Kalpakkam 603 102, Tamil Nadu, India

Dr. Venkataraman Balakrishnan

Department of Physics, Indian Institute of Technology
Madras 600 036, Tamil Nadu, India

Series Editors:
Professor Dr., Dres. h. c. Manuel Cardona
Professor Dr., Dr. h. c. Peter Fulde
Professor Dr., Dr. h. c. Klaus von Klitzing
Professor Dr. Hans-Joachim Queisser

Max-Planck-Institut für Festkörperforschung, Heisenbergstrasse 1
D-7000 Stuttgart 80, Fed. Rep. of Germany

Managing Editor:
Dr. Helmut K. V. Lotsch

Springer-Verlag, Tiergartenstrasse 17
D-6900 Heidelberg, Fed. Rep. of Germany

ISBN-13:978-3-642-83436-3   e-ISBN-13:978-3-642-83434-9
DOI: 10.1007/978-3-642-83434-9

Library of Congress Cataloging-in-Publication Data. Venkataraman, G. (Ganesan), 1932- Beyond the
crystalline state : an emerging perspective / G. Venkataraman, D. Sahoo, V. Balakrishnan. p. cm.–(Springer
series in solid-state sciences ; 84) Bibliography: p. Includes index. 1. Condensed matter. 2. Solid state
physics. I. Sahoo, D. (Debendranath), 1945-. II. Balakrishnan, V. (Venkataraman), 1943-. III. Title.
IV. Series. QC176.V46   1988   530.4'1–dc 19   88-12353

Typesetting: Macmillan India Ltd., India

2154/3150-543210 – Printed on acid-free paper

Preface

Condensed matter exhibits a rich variety of phases. Of these, the crystalline state has, until recently, received most attention. This is not surprising, given the geometric regularity of crystals. At the other extreme one has amorphous materials. In between there are the various types of liquid crystals, the recently discovered quasicrystals, and so on. While the absence of the high degree of regularity that characterizes the crystalline phase is certainly a problem, these noncrystalline states have nevertheless been receiving some attention over the years. However, it is only during the last few years that something like a unified view of all these phases has begun to emerge, through an application of various sophisticated concepts. Geometry and symmetry (and unusual realizations of the latter) provide a unifying thread in this new and emerging perspective.

This book is an attempt to capture the flavour of some of these recent developments. The approach is substantially descriptive, being intended to be accessible not only to experimental physicists, but also to chemists, materials scientists, metallurgists and ceramicists, whose work borders on physics. The prerequisites for a study of this book are a familiarity with basic solid-state physics and, in places, the elements of group theory and statistical mechanics. A few special topics are included at the end to aid those who wish to pursure further the subject matter treated here.

It must be emphasized that there is as yet no canonical viewpoint about the "grand unification" of the condensed state. What we present is therefore largely our own perception. The selection of topics naturally reflects our bias. Owing to limitations of space, we have restricted ourselves primarily to the *structural* aspects, touching upon properties only here and there. Indeed, many topics we should like to have included have regrettably had to be excluded. For these reasons, this volume should be regarded as a brief guide to some exciting current developments. If the book succeeds in some measure in arousing the curiosity of the reader and spurs him or her onto further pursuit of this emerging trend, it will have served its purpose.

We are indebted to M. C. Valsakumar for several discussions, critical comments and useful inputs at various stages. We would like to thank P. Subba Rao for his unstinted cooperation in typing the manuscript, and P. Harikumar for preparing the drawings. We are grateful to G. Athithan and S. Kingsley for the various computer graphics and to T. D. Sundarakshan, E. Viji, K. Neelakantan, V. Sridhar, A. K. Tyagi and B. Panigrahi for various forms of assistance. We also acknowledge the kind permission given by various publishers for the reproduction of material published earlier by them. One of us (GV) would like to thank the Jawaharlal Nehru Memorial Fund which made work on this book possible. He would also like to express his gratitude to Prof. J. S. King for his generous hospitality at the University of Michigan where some of this material was presented in the form of lectures. Finally, he is indebted to Mr. C. V. Sundaram, Director, Indira Gandhi Centre for Atomic Research, Kalpakkam, for personal encouragement and support.

July 1988

G. Venkataraman
D. Sahoo
V. Balakrishnan

Contents

1. Introduction

Plasma physicists are often fond of remarking that although over 99% of the Universe is a plasma, the subject of plasma physics remained neglected for a long time. There is a parallel of sorts in condensed matter physics: although the noncrystalline state is widely prevalent (especially in the living world), it is the crystalline state that has until now commanded the lion's share of attention from physicists. The reason is simply the geometric regularity of crystals which makes it *relatively* easy to analyze their properties. However, one has now begun to get an idea of the rules and logic Nature follows in forming the various noncrystalline states of matter, the crystalline state being a rather special case. The purpose of this volume is to offer a glimpse of some of these recent developments.

The level of the presentation is pedestrian, so as to make it accessible to a wide circle of readers, especially to nonphysicists. We consider this important because the noncrystalline states are often studied more extensively by nonphysicists than by physicists, and we believe that a better appreciation of current trends by these researchers could markedly stimulate further progress.

Chapter 2 is a brief description of the atomic structure of some of the important phases of condensed matter: crystals, incommensurate crystals, quasicrystals, liquid crystals and glass. In the case of systems other than crystals, it is neither convenient nor meaningful to describe the structure in terms of the coordinates of the individual atoms. More useful are the density–density and the orientational correlation functions, characterising respectively translational and orientational order. If the correlation function decays rapidly to zero as the distance increases, then the order is said to be of short range. On the other hand, if the correlation function tends to a finite, nonzero value as $r \to \infty$, long-range order exists. One can have short-range order alone (as in glass), or translational long-range order in one or more directions, or translational plus orientational long-range order, and so on. The various possible combinations lead to a rich variety of phases.

How does order emerge when a manifestly disordered system like a liquid is cooled, and how may this order be characterized? Chapter 3 provides some answers with the help of the concepts of symmetry breaking and the order parameter. We shall see that when a condensed state is formed, one or more of the symmetries of the parent disordered state are lost or "broken", and the emergence of order accompanies this lowering of symmetry. If the order is long ranged, then the correlation function of a suitable order parameter tends asymptotically to a constant value. Identification of the broken symmetry helps us understand the various possible combinations of short- and long-range translational and orientational order that occur.

Ordered states are not always perfect: for example, we know that dislocations occur commonly in crystals. Are the defects related in any way to the underlying symmetries of the ordered state? An arbitrary defect need not be so related, but *defects* intrinsic to the ordered system are. There is a subtle relationship between the symmetry of a system and the imperfections which can occur naturally in it. This relationship is explored in Chap. 4.

In Chap. 5 we consider the geometric description of some of the non-crystalline states. For crystals, such a description is straightforward in principle, but this is not so for noncrystalline states. There is as yet no prescription for a geometric description, but one strategy is to start with a suitable ordered reference structure (or "template"), and then "project" the desired structure out of it. The projection scheme is essentially a geometric algorithm. There are currently two projection schemes available; these are discussed and compared.

Any projection scheme is essentially a convenient method of generating a desired pattern, and there is no guarantee that a structure corresponding to the projected pattern actually exists. Considerations of energy and entropy enter, and these are explored in Chap. 6. The basic question is: What is the order-parameter configuration which corresponds to a free energy minimum, or at least a local minimum? This is an extension of the kind of question we deal with in introductory crystal physics: is it the fcc or hcp lattice of an assembly of argon atoms which is more stable at a given temperature and pressure?

The different structures one observes in condensed matter represent, in a sense, attempts to produce space-filling patterns using "building blocks" of various types, subject to certain optimization conditions on the free energy, entropy, etc. A very familiar space-filling problem is that of tiling the plane with identical regular pentagons without gaps or overlaps. It is well known that this is impossible. This exercise illustrates the difficulty frequently encountered in such problems: namely, that of meeting different requirements simultaneously (in this case, avoiding overlaps and voids on the one hand, and using only regular pentagons, on the other). The "generic" name for this conflict of interests is frustration, and Nature often seeks to relieve *frustration*, at least partially, via defects and disorder. (For example, in the instance above, if we are prepared to accommodate distorted pentagons as well as tetra-coordinated vertices, then a tiling with pentagons is possible.) The problems of frustration and its consequences have not yet been analyzed in their full generality. However, some recent studies are quite revealing, though restricted to a single spatial dimension. This is discussed in Chap. 7. We start with the work of Aubry on the arrangement of atoms along a chain when they are subjected to two periodic forces of different strengths and different wavelengths. A rich variety of structures is then possible. An interesting point about this approach is that it uses powerful methods from analytical dynamics; structures are classified in terms of orbits and trajectories in an appropriate phase space. If this approach can be extended to three dimensions, one could then ask: Given atoms of a particular kind, what are all the structures that one can build with them, both periodic as well as nonperiodic? It must be noted that this is quite different from the questions asked long ago by crystallographers, namely: What are the distinct types of periodic patterns one can have in Euclidean spaces of various dimensions?

Somewhat related to Aubry's work is that of Reichert and Schilling, who have studied spatially chaotic structures in one dimension as possible models for one-dimensional amorphous systems. This is also discussed in Chap. 7.

Symmetry breaking is only one of the causes for the occurrence of various phases of condensed matter. There appears to be a more comprehensive mechanism: ergodicity breaking. Among other effects, it seems to lead to the generation of certain atomic configurations by "freezing". Glass is a prime example. Chapter 8 deals briefly with ergodicity breaking. It appears to be necessary to consider this more general mechanism in order to understand the variety of structures in condensed matter.

In Chap. 9 we take a second look at all the concepts introduced earlier, and attempt to place them together in perspective.

The flow chart on the next page summarizes the sequence of concepts introduced and developed in this book.

A set of addenda on some special topics follows Chap. 9. The topics have been restricted to those that are directly relevant to the understanding, at a more detailed level, of certain concepts used in the main text. This set of topics is not meant to be exhaustive, nor is the treatment of any given topic complete: only the salient features have been presented. The intention is to make the book more self-contained in some measure, and also to aid the reader's progression to the original literature cited in the text.

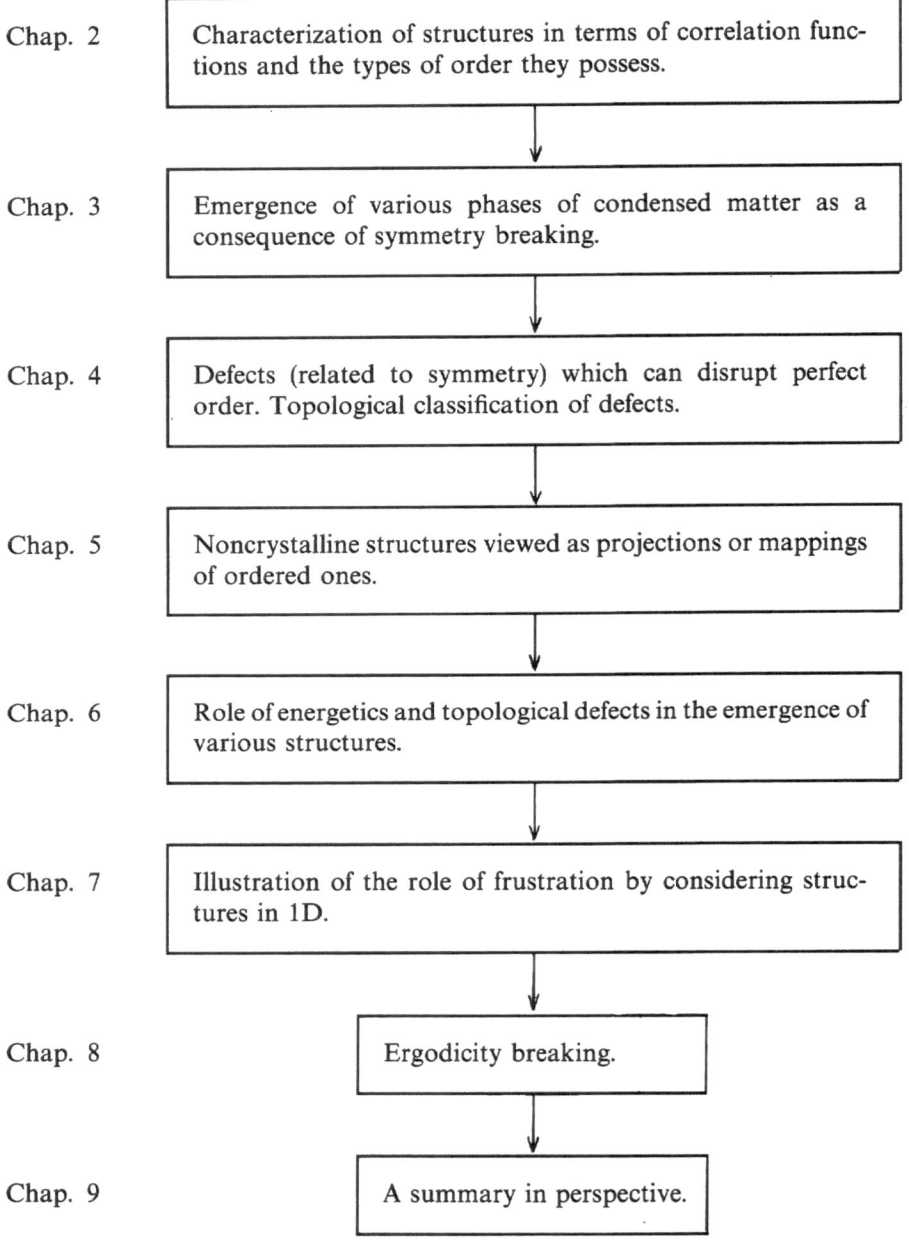

2. Variety in Structures

In this chapter we discuss the structure of some of the important phases of condensed matter, including the recently discovered quasicrystalline phase. The focus will be on the symmetry and geometry of the structure concerned, and on the positional correlations of the atoms in it.

2.1 Crystals

A crystàl represents a regular arrangement of atoms which can be generated by periodic translations of a basic motif or building block, the unit cell. The symmetry of a crystal is described by its space group. In three dimensions there are 230 space groups, representing the different possible combinations of point-group symmetry and translational symmetry. It is a standard result of crystallography that only certain rotational symmetries are consistent with translational periodicity. These are the so-called crystallographic point groups, of which there are 32.

A crystal has both symmetry and order. Order in the spatial arrangement of atoms is customarily characterized in terms of suitable correlation functions. Let $\psi(r)$ denote a quantity associated with the atomic arrangement at the point r. Then the correlation function

$$G_\psi(r) = \langle \psi(0)\psi(r) \rangle \tag{2.1}$$

describes the spatial correlations of ψ. Here $\langle \ldots \ldots \rangle$ denotes a thermodynamic average. In nonmagnetic systems, one is usually interested in the positional correlations of the atoms as described by

$$G(r) = \langle \varrho(0)\varrho(r) \rangle \tag{2.2}$$

where ϱ denotes the density. The function $G(r)$ is useful for exploring translational order.

Sometimes one is interested in the orientationally-averaged correlation function

$$G(r) = \frac{1}{4\pi}\int d\Omega \; G(r) \; . \tag{2.3}$$

Here $d\Omega$ is an element of solid angle in r-space. The quantity $G(r)$ is sometimes referred to as the *radial* correlation function. The *radial distribution function* is

$$P(r) = 4\pi r^2 G(r) \ . \tag{2.4}$$

In addition to translational order, one is often interested also in the orientational order. This is studied by means of the *orientational* correlation functions

$$G_n(r) = \langle Q_n(0)Q_n(r)\rangle \tag{2.5}$$

where $Q_n(r)$ is a suitably defined quantity characteristic of the interatomic bond centred at r (examples will be given below).

Long-range order (LRO) in ψ is said to exist when

$$\lim_{r \to \infty} \langle(\psi(0)-\langle\psi\rangle)(\psi(r)-\langle\psi\rangle)\rangle \neq 0 \ . \tag{2.6}$$

On the other hand, if the correlation function in (2.6) decays to zero as $r\to\infty$, then there is only short-range order (SRO). Crystals have both translational as well as orientational LRO. This statement is strictly true only in three dimensions (3D). In 2D the situation is somewhat different as we shall see in the next chapter.

2.2 Incommensurate and Long-Period Structures

A signature of the periodicity of a crystal is the periodicity of its diffraction spots. In recent years, many systems have been discovered for which the diffraction spots are accompanied by satellite reflections. It is now recognised that these satellites are due to certain periodic modulations which exist in the system. As a simple example, consider a 1D lattice (Fig. 2.1a) with sites

$$x_n = an \ , \tag{2.7}$$

where a is the lattice spacing and n is an integer. If the atoms are displaced to new positions X_n defined by

$$X_n = x_n + f\sin(qx_n) \qquad q = q_1(2\pi/a) \ , \tag{2.8}$$

then a modulated structure results. If q_1 is a nonzero rational number, then the structure (2.8) is also periodic, but with a larger unit cell (Fig. 2.1b). Such a structure is referred to as a *commensurate* or a *long-period* structure. When q_1 is irrational, there is no periodicity whatsoever, and one has an *incommensurate* structure (Fig. 2.1c). Although an incommensurate structure lacks periodicity, it has *quasiperiodicity*, a concept to which we shall return later.

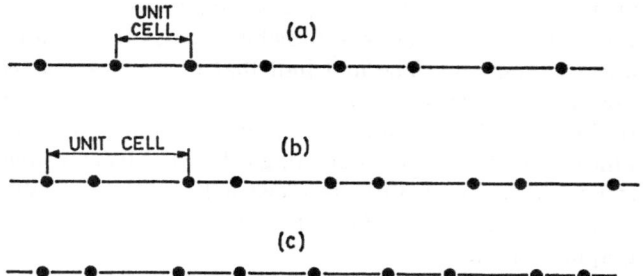

Fig. 2.1. (a) illustrates a periodic chain. A modulated structure results when the lattice sites are displaced as in (2.8). If the ratio of the periodicity of the modulation wave to that of the undisturbed lattice is a rational number, then the resulting structure is a commensurate one as in (b). Otherwise one obtains an incommensurate structure as in (c)

The 3D generalization of (2.7 and 8) is trivial, and is given by:

$$x(l, k) = x(l) + x(k) ,$$ (2.9a)

$$x(l) = l_1 a_1 + l_2 a_2 + l_3 a_3 ,$$ (2.9b)

where l_1, l_2, l_3 are integers, a_1, a_2, a_3 are the basis vectors of the crystal lattice, $x(k)$ denotes the position of the kth atom in the unit cell ($k = 1, \ldots, s$), and

$$X(l, k) = x(l, k) + f \exp[i q \cdot x(l)] .$$ (2.10)

Incommensuration can also be non-displacive in nature. For example, it may arise as a result of magnetic modulation (as in some rare earths), or compositional fluctuations (also called substitutional modulation) in non-stoichiometric systems [2.1].

2.3 Quasicrystals

Late in 1984, a new kind of quasiperiodic system was discovered in rapidly cooled Al–Mn alloys, creating considerable excitement [2.2]. Referred to as a *quasicrystal*, it is essentially a quasiperiodic system, but with a *noncrystallographic* orientational symmetry. By contrast, an incommensurate crystal is a quasiperiodic system possessing a crystallographic orientational symmetry, i.e., its rotational symmetry is given by one of the 32 crystallographic point groups (in 3D). These statements will be amplified later.

We have already remarked that only certain rotational symmetries are consistent with translational periodicity. As a result, it is, for example, not possible to tile the plane with identical regular pentagons without having either voids or overlaps. Since periodicity implies LRO, it was assumed (until recently) that no structure with translational LRO was possible, possessing a noncrystallographic orientational symmetry. The first hint that this might not be true came from some recreational studies by *Penrose* [2.3] who was interested in filling the plane nonperiodically with the minimum number of tile shapes. He discovered that a pair of tile shapes plus a set of matching rules (which determine how adjacent tiles are to be joined) enabled one to construct such a tiling. We now know that quasiperiodic translational LRO is also possible and that it can exist with noncrystallographic symmetry.

Figure 2.2 is an example of a Penrose "lattice" composed entirely of two rhombic unit cells. This pattern has both five-fold orientational symmetry and quasiperiodicity in a sense to be described. For this reason, the Penrose tiling may be regarded as a 2D example of a quasicrystal. The orientational symmetry and the quasiperiodicity bestow the Penrose tiling with both long-range orientational as well as translational order.

Like a crystal, a quasicrystal also has a repeating motif, i.e., it too can be regarded as built up from unit cells, but more than one in number. In the example of Fig. 2.2, there are two unit cells. Unlike the case of crystals, however, we do not have a set of group-theoretic transformations to generate the entire structure, starting with the basic unit cells. But there are other ways of generating quasiperiodic lattices. Before discussing them, we must first clarify what is meant by quasiperiodicity.

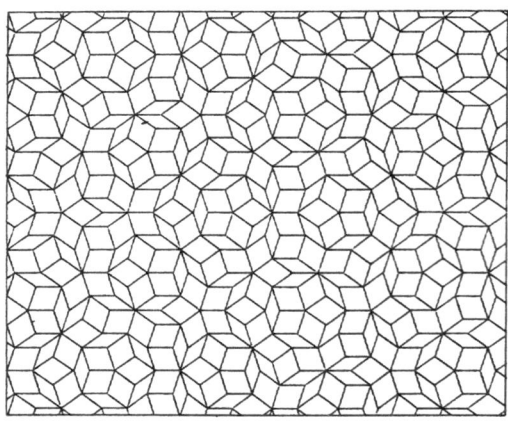

Fig. 2.2. Example of a Penrose tiling. It is composed entirely of two types of rhombuses ("fat" and "thin"). The tiling is not periodic but quasiperiodic

2.3.1 Quasiperiodicity

Roughly speaking, a quasiperiodic function is a superposition of periodic functions whose periods are incommensurate with respect to each other. A simple example would be the density function

$$\varrho(x) = \cos qx + \cos Qx , \quad (Q/q) \text{ irrational} . \qquad (2.11)$$

Our immediate interest, however, is in a quasiperiodic sequence of points. Consider the 1D chain in Fig. 2.3a. Let the step length be S. Introduce another (long) step L; the value of the ratio (L/S) will be considered later. From the chain of Fig. 2.3a, we generate a series of new chains using two replacement rules, namely, (i) $S \to L$, and (ii) $L \to LS$. By applying these on the lattice in Fig. 2.3a, we obtain first the lattice in Fig. 2.3b, and then the lattice in Fig. 2.3c, and so on. Except for the first two lattices, all others feature two tile shapes; only the length of the period goes on increasing.

Table 2.1 summarizes the evolution. Observe that the periodicity increase follows the Fibonacci sequence determined by the recursion formula

$$f_0 = 1 , \quad f_1 = 1 , \quad f_k + f_{k+1} = f_{k+2} . \qquad (2.12)$$

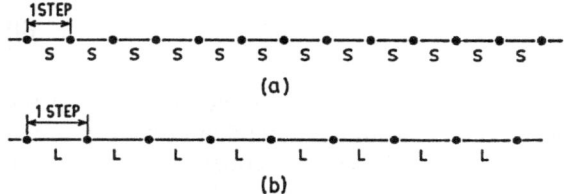

(a)

(b)

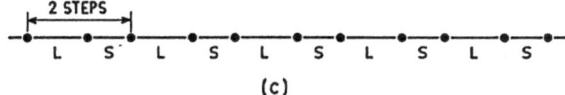

(c)

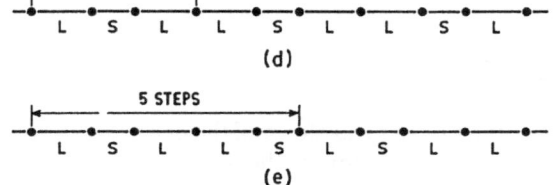

(d)

(e)

Fig. 2.3a–e. Illustration of how a quasiperiodic sequence involving two types of tiles can be obtained using the replacement rules discussed in the text

Table 2.1. Evolution of the Fibonacci lattice

Period		1	1	2	3	5	8	13	21	...	
$\dfrac{(\text{No. of L})}{(\text{No. of S})}$				$\frac{1}{1}$	$\frac{2}{1}$	$\frac{3}{2}$	$\frac{5}{3}$	$\frac{8}{5}$	$\frac{13}{8}$	...	1.6182 ...

It is well known that

$$\lim_{k \to \infty} (f_{k+1}/f_k) = \tau = (1+\sqrt{5})/2 \quad \text{(the golden mean)} .$$

Further, the stoichiometry, i.e., the ratio of the number of L tiles to the number of S tiles, also tends to the golden mean.

Consider the infinite-period Fibonacci sequence . . . $LSLLS$ . . . derived as discussed above. Suppose *further* that the tile lengths are given by $S = 1$ and $L = \tau$. Starting the sequence as $SLSLLSL$ . . . . . , the position of the Nth point may be expressed as [2.4, 5]

$$x_N = N + \frac{1}{\tau} \left\lfloor \frac{N}{\tau} \right\rfloor , \tag{2.13}$$

where $\lfloor x \rfloor$ denotes the "floor" of x (the largest integer less than or equal to x). This may be generalized to

$$x_N = N + \alpha + \frac{1}{\tau} \left\lfloor \frac{N}{\tau} + \beta \right\rfloor , \tag{2.14}$$

where α and β are arbitrary real numbers. α determines the origin and β the sequencing of the tiles. Thus two sequences with the same β but different values of α are related by a mere shift of the origin, whereas the sequences with the same α but different values of β differ in the ordering of the tiles. A chain with lengths L and S as above will be referred to as a *Penrose chain* or a 1D *Penrose lattice*. The density function for the chain is

$$\varrho(x) = \sum_N \delta(x - x_N) . \tag{2.15a}$$

Let us choose

$$x_N = \frac{1}{\sqrt{1+\tau^2}} \left(N + \frac{1}{\tau} \left\lfloor \frac{N}{\tau} + \frac{1}{2} \right\rfloor \right) . \tag{2.15b}$$

It is interesting to examine $\varrho(q)$, the Fourier transform of the density function (see also Sect. 5.17). This is given by an expression of the form [2.4, 6]:

$$\varrho(q) = \sum_{m,n} A_{mn} \delta(q - q_{mn}) . \tag{2.16}$$

The coefficients A_{mn} are worked out later and given precisely by (5.12, 14); and

$$q_{mn} = \frac{2\pi}{\sqrt{1+\tau^2}} (m + n\tau) , \quad m, n \text{ integers} . \tag{2.17}$$

The point is that $\varrho(x)$ is a superposition of density waves with wavevectors $\{q_{mn}\}$ which, on account of (2.17), are incommensurate with each other. The Penrose lattice is consequently quasiperiodic.

2.3.2 2D Quasicrystalline Tilings

Turning to 2D, the construction of a quasiperiodic network starts with a specification of the orientational symmetry one is aiming at. This is best illustrated by considering the case of five-fold symmetry, especially since it is directly related to the Penrose tiling. To understand the method, let us first consider the simpler example of the generation of a square lattice. We start with a star of four vectors at 90° to each other. Actually, on account of symmetry, it is sufficient to just consider a pair at right angles to each other. Periodically spaced lines (unit distance apart, say) are now drawn normal to the two chosen members of the star. The set of intersections of the two sets of parallel lines yields the desired square lattice. For generating Penrose tilings, *de Bruijn* [2.7] has suggested a similar method, involving five periodic grids rotated by angles of 72° and its multiples with respect to each other (Fig. 2.4a). To start with, one line in each fringe system is arranged to pass through the origin. The grids are then given lateral displacements γ_i such that $\sum_i \gamma_i = 0$. The grids now intersect as in Fig. 2.4b. (It is necessary to adjust the $\{\gamma_i\}$ so that more than two lines do not

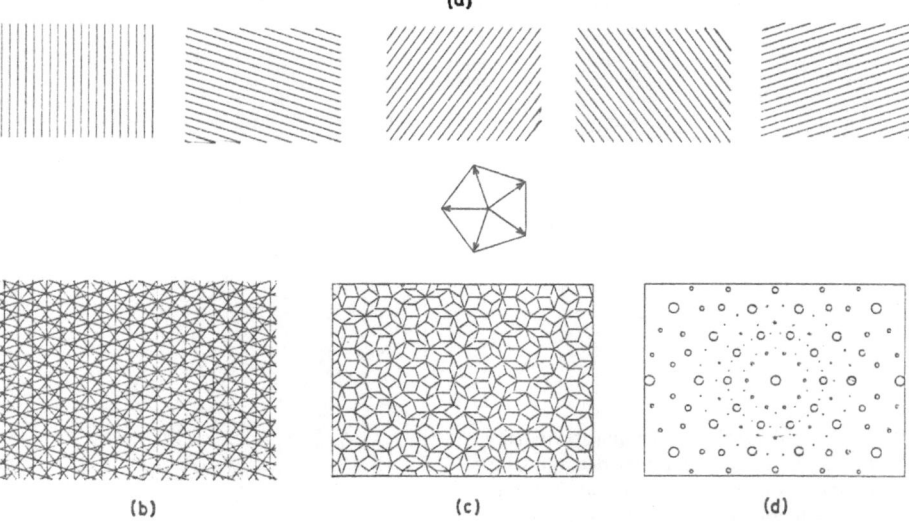

Fig. 2.4. (a) shows five grid systems normal to the directions of five corners of a regular pentagon. A typical intersection of these grids is as in (b). Upon decoration as described in reference [2.7] there results the Penrose tiling (c). The five-fold symmetry of the tiling is evident from the diffraction pattern (d) where the (relative) intensity is proportional to the diameter of the circle corresponding to the diffraction spot

intersect at a point. This would represent a singular situation which is to be avoided). Each intersection is then decorated according to a given prescription [2.7], whereupon a Penrose tiling results (Fig. 2.4c).

Looking at the tiling, two questions arise: (i) Where is the five-fold symmetry, and (ii) how does the tiling exhibit quasiperiodicity? The orientational symmetry of the Penrose tiling is somewhat subtle. While a large variety of tilings is possible (these variations can be achieved by choosing the set $\{\gamma_i\}$ differently, subject to the sum rule $\sum_i \gamma_i = 0$), only in exceptional cases does the pattern behave like a giant molecule with a pentagonal point group symmetry. The orientational symmetry of an *arbitrary* Penrose tiling cannot, however, be so described[1]. Nevertheless, there is orientational order in the sense that every bond in the tiling points in one or the other of the five directions defined by the normals to the five grids. It turns out that although the orientational symmetry of the direct lattice is not so transparent, it becomes immediately evident when one examines the diffraction pattern (see, for example, Fig. 2.3d). In fact, the diffraction pattern also has the usual inversion symmetry.

The demonstration of quasiperiodicity is also subtle. Using a suggestion due to Ammann, *Levine* and *Steinhardt* [2.4] have decorated the tiles of the Penrose tiling as in Fig. 2.5a. The line segments then join to form continuous lines, leading to five grid systems (Fig. 2.5b). The spacings between the lines follow those of a Penrose chain.

Socolar et al. [2.8] (see also [2.5]) have extended de Bruijn's method to what they call the *generalized dual method* (GDM). The procedure in 2D is as follows:

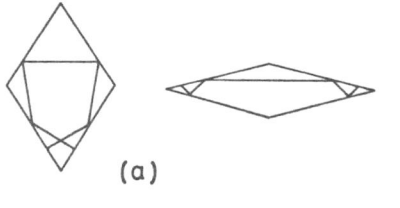

(a)

(b)

Fig. 2.5. To demonstrate the quasiperiodicity of the Penrose tiling, one first decorates the basic tiles with segments as in (**a**). The segments then join to form five quasiperiodic grids as in (**b**)

---

[1] Many quasicrystals (including Penrose tilings) may be assigned a symmetry group in a higher dimensional space. We shall discuss this in Chap. 5.

(i) A star of N vectors is first chosen which determines the desired orientational symmetry. (ii) Quasiperiodically spaced lines are next drawn normal to each star vector e_i and numbered. For instance, the spacings could follow (2.13) (which is a special case). Each set of parallel lines so drawn is referred to as a *grid*, and all grids taken together form an *N-grid*. The grid lines are drawn so as to avoid singular intersections (as mentioned earlier). (iii) The grid lines divide the plane into nonintersecting, open regions through which no lines pass. Let x_0 be a point in one such region, and let it lie between lines r_i and $(r_i + 1)$ which are normal to e_i. An integer r_i is then assigned to x_0. In this way, by considering all the N different grids, each open region can be specified by N integers $(r_1, \ldots, r_N)$. (iv) For each integer set $(r_1, \ldots, r_N)$, one constructs a vector $t = \sum_i r_i e_i$. The points t define the vertices of a quasilattice with the orientational symmetry of the initial star. By joining neighbouring points suitably, one can obtain the desired tiling pattern. If the star consists of five vectors pointing from the centre to the five vertices of a regular pentagon, and the spacings of each grid follow the Fibonacci sequence (2.13), then the resulting quasilattice is nothing but the Penrose tiling.

The set of points $\{t\}$ form a *dual* to the set of intersection points of the N-grid. One may wonder whether the required nonperiodic tiling cannot be generated using the intersection points themselves. Indeed it can be, by constructing the Voronoi polyhedron (or the Wigner–Seitz cell) around each intersection point. One then obtains a space-filling tiling with the desired orientational symmetry but having many more unit cells. The dual method is more economical [2.5].

2.3.3 A Brief Recapitulation

So far we have been discussing abstract quasiperiodic *lattices*, which are really like the Bravais lattices of crystallography. Physical quasicrystalline structures result when the lattices are decorated suitably with atoms and molecules [2.9]. The essential feature is, of course, quasiperiodicity.

The Penrose chain is a special example of a quasiperiodic chain with two unit cells L and S, that is generated via the repeated application of the substitution rule

$$\begin{pmatrix} L \\ S \end{pmatrix} \rightarrow \begin{pmatrix} 1 & 1 \\ 1 & 0 \end{pmatrix} \begin{pmatrix} L \\ S \end{pmatrix} \qquad (2.18)$$

with the ratio of lengths $(L/S) = \tau$. This leads to the sequence (2.14). A slightly more general quasi-periodic sequence has the form

$$x_N = a \left(N + \alpha + \frac{1}{\varrho} \left\lfloor \frac{N}{\sigma} + \beta \right\rfloor \right) ; \quad \sigma, \varrho > 0 , \qquad (2.19)$$

where a is some length scale. Here the long and the short tiles have lengths $a(1 + \varrho)/\varrho$ and a respectively. The stoichiometry is given by (No. of L)/(No. of S) $= 1/(\sigma - 1)$.

One need not be restricted to using just two tiles. In general, a k-quasicrystal (in 1D) can be generated via the substitution rule

$$L_i \rightarrow \sum_j M_{ij} L_j \ . \tag{2.20}$$

Here the matrix $\underline{M}$ is a $(k \times k)$ non-singular matrix with non-negative integer elements whose eigenvalues satisfy a polynomial equation of the kth degree. Further generalizations are also possible.

The 2D Penrose tiling is a special kind of quasiperiodic lattice. It is also *self-similar*. A simple example in 1D is shown in Fig. 2.6. Here, by decorating the tiles suitably, one is able to obtain a new tiling with scaled-down unit cells. This process of reducing the lattice to one that is similar in form is called *deflation*. Alternately, one may systematically omit some vertices to obtain an enlarged but identical Penrose tiling (*inflation*). The quasiperiodic lattices generated by the GDM are not necessarily self-similar.

Earlier, while discussing the generation of Penrose tilings, we noted that by changing the set $\{\gamma_i\}$ different patterns can be generated, all of them featuring the same pair of rhombic unit cells. Are these different patterns related to each other? In general, two arbitrary patterns need not be related to each other. However, the set of all possible tilings can be grouped into distinct classes such that the tilings belonging to any one class are *locally isomorphic*. Two quasi-crystals are in the same local isomorphism class if and only if every finite configuration of unit cells that appears in one also appears in the other. More simply, if one draws a circle of diameter d about a lattice point P in one quasicrystal, the pattern of unit cells contained within the circle must be found in the other quasicrystal for it to be locally isomorphic to the former, and this must be true for all finite d and all choices of P.

Local isomorphism has considerable physical significance [2.4]; for instance, two quasicrystals in the same local isomorphism class have the same free energy. It is also clear that, owing to the high degree of degeneracy implied by the

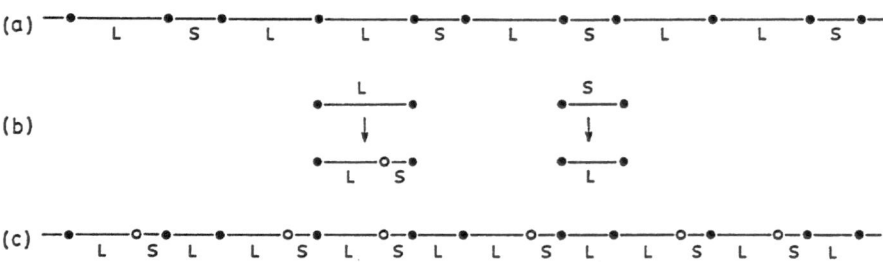

Fig. 2.6. (a) shows a 1D tiling following a Fibonacci sequence. The tiles are now replaced with smaller tiles as illustrated in (b). The resulting chain (c) is a reduced copy of (a)

existence of a local isomorphism class, there can be no unique pattern generation technique starting with the basic unit cells, unlike the case of crystals.

2.3.4 3D Quasicrystals

2D quasicrystals are interesting in the context of some materials which show quasiperiodicity in the plane, but conventional periodicity along a direction perpendicular to the plane (the decagonal phase) [2.10]. On the other hand, there is an increasing number of examples of materials that exhibit 3D quasi-periodicity.

The 3D case is somewhat more involved [2.5], but the basic rules for generating the structure are similar to those discussed earlier. Here we shall restrict our attention to the so-called icosahedral quasicrystals. To generate these, one starts with a set of six unit vectors directed along the six five-fold symmetry axes of an icosahedron (Fig. 2.7). Planes are drawn normal to each of these vectors e_i ($i = 1, \ldots, 6$), each set of planes having a Fibonacci sequence of spacings. In this way, one obtains a *Fibonacci hexagrid*. One adjusts the grids such that any triplet of planes belonging to the ith, kth and the jth grids (where $i \neq j \neq k$) intersect exactly at one point. Thus, contrary to the situation in 2D, it is the *singular* grid system which becomes physically relevant. The planes divide the grid space into open regions. Duals of these regions are formed, as before, by constructing vectors $t = \sum_i r_i e_i$, where the integer sets $(r_1, \ldots, r_6)$ label the open regions. A quasiperiodic 3D network then results. This may then be decorated with atoms to obtain prototypes of realistic structures. It should be noted that this construction gives the analogue of Penrose tilings in 2D, i.e., structures which are also self-similar. The set of all possible 3D quasicrystals is larger than this set.

The structure, stability and physical properties of quasicrystals are receiving considerable attention at the moment. We refer the reader to the proceedings of a recent Workshop on quasicrystals [2.11] to get an idea of the status of this rapidly expanding field.

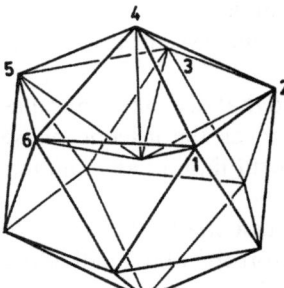

Fig. 2.7. Set of six unit vectors needed for generating the hexagrid. The vectors are directed along the five-fold symmetry axes of an icosahedron

2.4 Liquid Crystals

Many organic crystals melt by passing through intermediate fluid phases that are optically anisotropic. Discovered in the last century, the state of matter in these intermediate phases is now referred to as *liquid crystalline* for reasons which will become clear shortly. A more technical name is *mesomorphic* phase (mesomorphic: of intermediate form).

Liquid crystals are *broadly* classified as *nematics, cholesterics* and *smectics* [2.12, 13]. To understand why the liquid crystal phase is an "intermediate" one, we note first that a crystal has both long-range translational as well as orientational order. On the other hand, in a liquid both these are destroyed. In the mesomorphic phases one has an in-between situation which can be realized in a variety of ways. Such phases become likely in systems composed of anisotropic building blocks. The classic example of such a molecule is p-azoxyanizole (PAA) with the formula

$$CH_3-O-\langle\!\!\!\bigcirc\!\!\!\rangle-\overset{\displaystyle N=N}{\underset{\displaystyle O}{|}}-\langle\!\!\!\bigcirc\!\!\!\rangle-O-CH_3.$$

Liquid crystals can also be built up from helical and disc-like molecules.

The simplest of the various liquid crystalline phases is the nematic, illustrated in Fig. 2.8a. Here the molecules are thread-like and aligned parallel to each other on the average. However, the centres of gravity of the molecules have no LRO. In short, there is long-range orientational order but only short-range positional or translational order.

The cholesteric phase illustrated in Fig. 2.8b is formed when chiral (helical) molecules are present. Locally, a cholesteric is rather similar to a nematic. The molecular axes are all aligned, but the centres of gravity are disordered. The local alignment may be characterized by a quantity n called the *director* which specifies, roughly speaking, the average direction along which the rod-like molecules are aligned. Globally, n is not constant in a cholesteric; it changes direction, spiralling around an axis.

Smectics occur in various varieties, of which one example is given in Fig. 2.8c. There being a reasonably well-defined layer structure, this system clearly has greater order as compared to the nematic.

Recently *Chandrasekhar* et al. [2.14] have discovered that disc-like molecules can also give rise to a mesomorphic phase referred to as the *discotic* phase (Fig. 2.8d). The molecules are irregularly stacked to form liquid-like columns. The columns themselves are closely packed.

The anisotropy of the liquid crystalline phase is essentially the result of orientational order. Such order is facilitated by the anisotropy of the individual building block. It may be asked whether such an intermediate phase is possible in

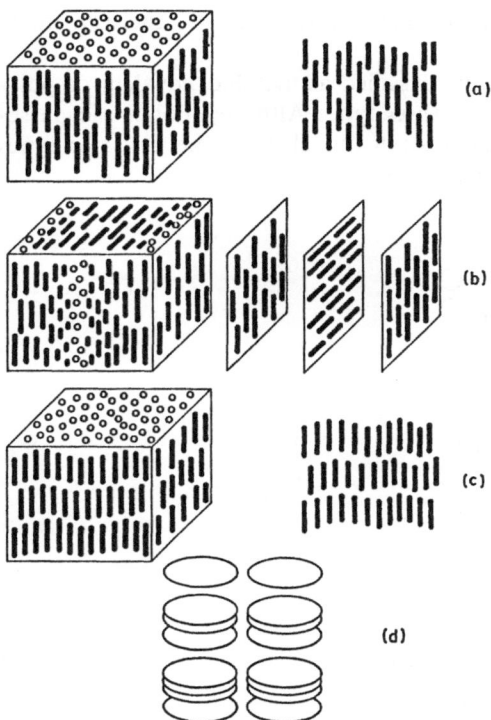

Fig. 2.8a–d. Typical liquid crystalline phases. The nematic shown in (**a**) has only orientational order. (**b**) shows the cholesteric phase. Here the molecules have the same alignment in local regions. ·On a more extended scale, the alignment spirals periodically. The smectic A is shown in (**c**). The molecules tend to lie in planes with no configurational order within planes, and to be oriented perpendicular to the planes. It is customary to idealize the stacking of planes by perfect periodicity. The discotic phase is illustrated in (**d**)

a hard sphere system (or, say, in argon). In this case there is no preferred molecular axis to aid orientational order. Nevertheless, one can conceive of orientational order in the system in terms of near-neighbour bonds, as discussed earlier with regard to quasicrystals. The question is: Can there exist, for a system of spheres (both hard and soft), a phase with no long-range translational order but with long-range bond-orientational order? This would be the analogue of the liquid crystalline phase with atoms as the building blocks rather than anisotropic molecules. An example of such a phase in 2D is the hexatic phase suggested by *Halperin* and *Nelson* [2.15]. Unfortunately, there is no conclusive experimental proof that this phase exists [2.16]. In 3D, there are no theoretical grounds as yet for expecting such a phase.

2.5 Glass

Broadly speaking, a glass is a congealed melt, rigid but devoid of crystalline order. If a crystal is a paradigm of order then glass is just the antithesis. The glass we are most familiar with is one example of an amorphous solid; there are many others.

Amorphous solids may be roughly classified as metallic, covalent or polymeric [2.17]. Figure 2.9 illustrates these three types. Although polymeric glasses are important, in this book we shall restrict ourselves to the first two types whenever we consider amorphous structures.

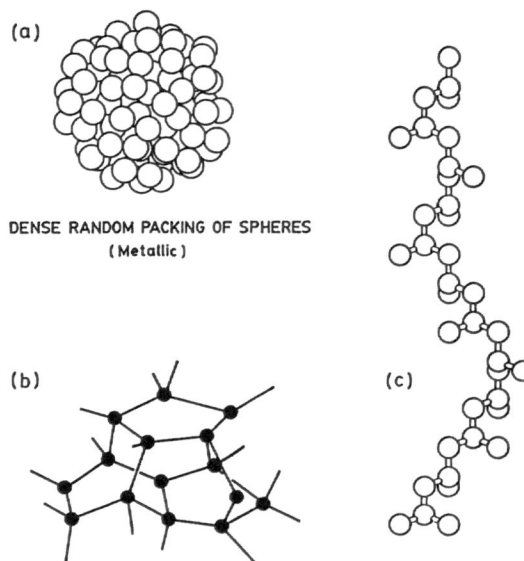

(a)

DENSE RANDOM PACKING OF SPHERES
(Metallic)

(b)

CONTINUOUS RANDOM NETWORK
(Covalent)

(c)

Polymeric

Fig. 2.9a–c. The three broad types of amorphous materials: (a) sphere-packed, (b) covalently-bonded and (c) polymeric

Atomic arrangements in amorphous materials are sometimes referred to as random. This is not strictly correct. If, for example, we try to build a disordered assembly of spheres of radii R by taking one sphere as a nucleus and packing others around it, then it is clear that the centre of no other sphere can approach that of the initial sphere closer than a distance $2R$. This exclusion automatically produces some correlations, but of very short range. In amorphous materials, the range over which translational and orientational correlations decay to zero is *finite*.

There are two important prototype structures for amorphous solids of the metallic and covalent types. These are the *dense random packing* (DRP) and the *continuous random network* (CRN) respectively. The DRP concept owes much to *Bernal* [2.18, 19] who viewed a liquid as a *heap* of atoms instead of a *regular pile* which is what a crystal is. Bernal evolved clever ways of building DRP's by hand (a task that is much harder than it looks); today the problem is amenable to computer simulation [2.20, 21]. An important outcome of Bernal's work is that tetrahedral clusters of spheres play a prominent role in DRP structures.

The CRN idea goes back to several workers [2.22–24] who, like Bernal, tried to build models of tetracoordinated structures by hand. In effect, a CRN is a space-filling network built of tetrapods (Fig. 2.9) with no dangling bonds. Randomness enters in the way adjacent tetrapods are linked, i.e., in the dihedral angle (Fig. 2.10). Some minor fluctuations in bond lengths also occur on account of the need to bend some bonds. In the final network, every vertex is tetracoordinated. There is thus a short-range correlation as in the DRP.

The DRP and the CRN are closely related [2.25]. This is understandable, because the building blocks of the two, the tetrahedral cluster and the tetrapod, are "duals" of each other (Fig. 2.11).

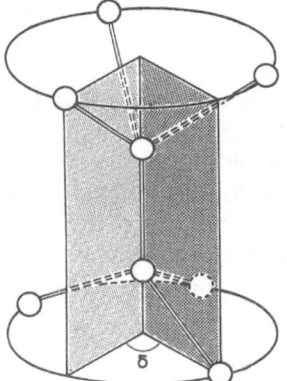

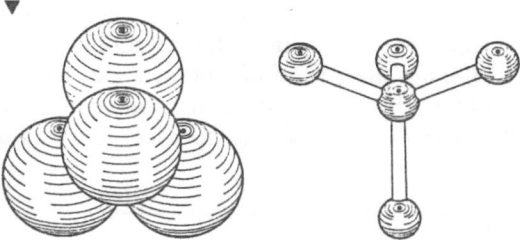

Fig. 2.11. Building blocks of the DRP and the CRN structures

Fig. 2.10. Dihedral angle δ between adjacent tetrapods. In a CRN structure, this angle varies randomly

2.6 Systems with Quasi Long-Range Order

We have already explained what is meant by LRO. *Short*-range order is generally characterized by the exponential decay

$$G_\psi(r) \sim \exp(-r/\xi_\psi) \,, \tag{2.21}$$

where ξ_ψ is the *correlation length*, usually of the order of a few interatomic spacings. There are, however, situations where the correlations decay much more slowly, according to a power law

$$G_\psi(r) \sim r^{-\eta} \,. \tag{2.22}$$

In these cases, the behaviour of G_ψ is intermediate between the standard exponential decay and true LRO. Such behaviour is sometimes referred to as quasi-LRO [2.26]. An example is the behaviour of translational correlations in a 2D crystal [2.16]. This will be discussed further in the next chapter.

2.7 Overview

In this chapter we have been concerned mainly with the different ways in which atoms and molecules are arranged in space in various phases of condensed matter. All structures can be characterized in terms of the correlation functions $G(r)$ and $G_n(r)$. Different phases can be distinguished (to some degree) by the extent to which translational and orientational correlations prevail. Some of these distinctions are illustrated in Fig. 2.12 which compares the radial distribution functions for a 2D hexagonal lattice, a Penrose tiling and a 2D hard disc gas. A more comprehensive summary is available in Table 2.2.

Of all the structures possible, the crystalline one is unique in that the extended structure can be built up systematically and in a *unique* manner by starting with the unit cell and applying operations deducible from symmetry. In the quasicrystal, some of the crystalline features survive (like the orientational LRO), but there is no unique structure resulting from the repetition of the basic motif. In this respect, the situation is somewhat like that in an amorphous solid. The liquid crystal phases also lack the uniqueness that is characteristic of the

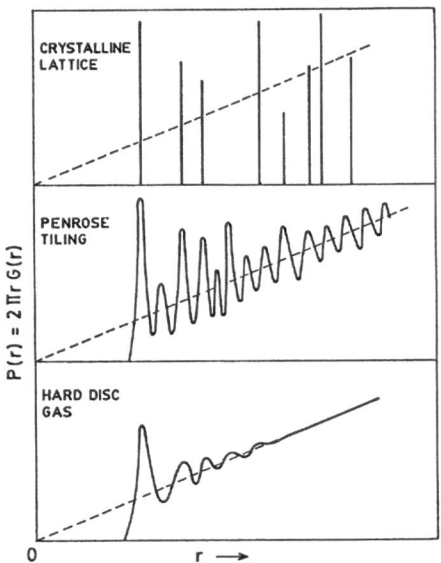

Fig. 2.12. Radial distribution functions (schematic) for a 2D hexagonal lattice, a Penrose tiling and a 2D hard disc gas. In the first two cases, the correlations are long-ranged

Table 2.2. Summary of the state of order in some condensed matter systems

Translational order	Orientational order	Systems
LRO	LRO	Crystals, incommensurate crystals, quasicrystals
SRO	LRO	Liquid crystals[a]
LRO	SRO	Plastic crystals[b]
SRO	SRO	Liquids, Amorphous solids

[a] Depending on the liquid crystal one is considering, one could have a coexistence of SRO and LRO. For example, smectic A is liquid-like along two directions and crystal-like in the third, as far as translational order is concerned. The classification we have indicated is thus a bit gross (although it applies exactly to nematics).
[b] Plastic crystals are usually composed of nearly spherical molecules, in contrast to liquid crystals whose building blocks are linear. In the intermediate (i.e. plastic crystalline) phase there is translational order but no orientational order.

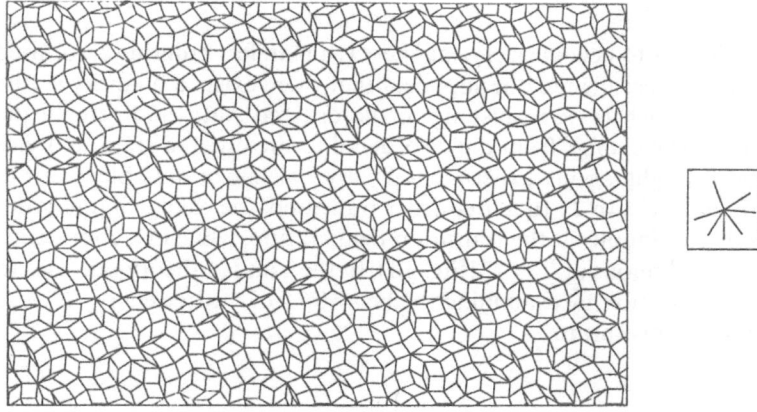

Fig. 2.13. 2D tiling generated using the method discussed in Sect. 2.2.2, for an arbitrary star of the type shown in the inset. There are many types of unit cells, and the translational quasiperiodicity tends to get masked. Correspondingly, the diffraction peaks tend to merge, forming rings similar to what one observes in an amorphous structure. However, there is bond orientational order, i.e., each bond points along one of the directions of the star

crystalline structure. Of course in amorphous structures the disorder is even more manifest.

Sometimes, the distinction between different phases can be blurred as illustrated in Fig. 2.13. This shows a 2D quasiperiodic system with arbitrary orientational symmetry. In this case there are many unit cells, and it is not easy to distinguish the pattern from a 2D amorphous structure.

Why are there so many different phases, and how do they arise when a liquid is cooled? Some answers are discussed in the next chapter.

3. Order Out of Disorder

This chapter is concerned with the emergence of order out of disorder, and specifically with symmetry breaking in phase transitions in the framework of the Landau theory [3.1, 2].

3.1 Landau Theory

When a liquid freezes to become a crystal, there is a manifest ordering associated with the process. Generalizing, one supposes that there exists a quantity ψ (the *order parameter*) which is nonvanishing in the ordered state, and zero in the disordered state. The change from the disordered to the ordered state, and vice versa, can be brought about experimentally by varying a suitable thermo-dynamic variable like T or P. In a first-order transition, the change in the order parameter is discontinuous at the transition temperature, while in a continuous transition, ψ is continuous at $T = T_c$ (Fig. 3.1). In the Landau theory one tries to understand the transition by examining the behaviour of the free energy F close to the transition temperature.

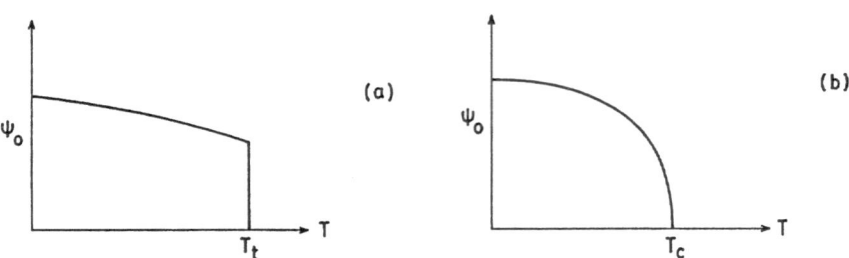

Fig. 3.1. Variation of the order parameter ψ_0 with temperature in a first order (**a**) and second order (**b**) transition. The temperatures T_t and T_c denote the transition temperatures in the two cases

3.1.1 Transition in a System with a Scalar Order Parameter

The order parameter can in general be a complicated quantity (as we shall see later). Here we suppose that it is a real scalar (i.e., it has only one component), and further that the free energy is insensitive to the sign of the scalar. F can then be expressed as

$$F(\psi) = F_0 + \alpha\psi^2 + \beta\psi^4 \tag{3.1}$$

in the vicinity of the transition temperature T_c, where F_0 is a reference level (the value of F at T_c), and β is a positive constant. The coefficient α is a smooth function of T in the neighbourhood of T_c, with a leading behaviour

$$\alpha \simeq a(T - T_c) , \quad a > 0 . \tag{3.2}$$

Thus α is positive for $T > T_c$ while it is negative for $T < T_c$. Figure 3.2 shows a sketch of F corresponding to different temperatures in the neighbourhood of T_c.

The thermodynamically stable state ψ_0 of the system is found by minimizing F with respect to ψ. The stability conditions are

$$(\partial F/\partial\psi)_{\psi_0} = 0 , \quad (\partial^2 F/\partial\psi^2)_{\psi_0} > 0 . \tag{3.3}$$

A simple calculation then shows that for $T > T_c$, $\psi_0 = 0$ is the stable state, whereas for $T < T_c$, stable states with non-vanishing values for the order parameter are possible, as Fig. 3.2 illustrates qualitatively.

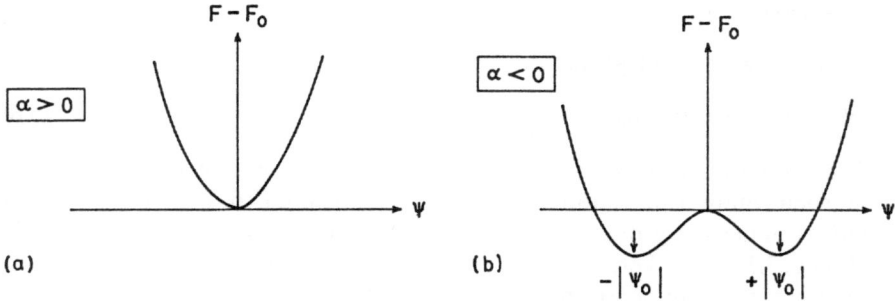

Fig. 3.2a, b. Schematic plots of the free energy function in (3.1) for $T > T_c$ and $T < T_c$

3.1.2 Role of Symmetry

Symmetry plays a key role in many phase transitions, and the example just discussed provides an illustration. The free energy (3.1) is invariant under the transformations $\psi \to \psi$ and $\psi \to -\psi$, i.e., under the discrete two-element group $\mathbb{Z}_2 = \{E, I\}$ where E is the identity transformation and I is the inversion. Labelling the thermodynamic state by its order parameter value ψ_0, we find that, for $T > T_c$,

$$E\psi_0 = \psi_0 , \quad I\psi_0 = \psi_0 , \tag{3.4}$$

since $\psi_0 = 0$ in this case. For the nontrivial stable states $\pm \psi_0$, below T_c,

$$E(+\psi_0) = (+\psi_0) \ , \qquad E(-\psi_0) = (-\psi_0) \ , \tag{3.5a}$$

$$I(+\psi_0) = (-\psi_0) \ , \qquad I(-\psi_0) = (+\psi_0) \ . \tag{3.5b}$$

From (3.5b) we see that whichever one of two states is selected, it does *not* have the symmetry of the group $\mathbb{Z}_2$. Thus, below T_c, the thermodynamic state lacks the full symmetry of F, and we have a state of *broken symmetry*. In general, symmetry is said to be broken when the thermodynamic state does not exhibit the same symmetry as is present in the free energy. Symmetry breaking and the appearance of order are thus closely related.

3.1.3 Systems with a Complex Order Parameter

Consider next the case where ψ is a complex order parameter (and therefore has two real components). We now have the expansion

$$F = F_0 + \alpha |\psi|^2 + \beta |\psi|^4 \ , \qquad \beta > 0 \ . \tag{3.6}$$

Since the order parameter is complex,

$$\psi = |\psi| \exp(i\theta) \ . \tag{3.7}$$

Consider now the transformation

$$\psi \rightarrow \psi' = [\exp(i\omega)]\psi \ , \qquad (\omega \text{ real}) \ . \tag{3.8}$$

This corresponds to a rotation by the angle ω in the complex plane. The set of all such rotations constitutes the group U(1). From (3.6) it is clear that F is invariant under the group of transformations U(1).

Figure 3.3 shows sketches of the free energy surface for $T > T_c$ and $T < T_c$. In the low-temperature phase the minima of F lie on a circle. Every point on the circle represents a possible ordered state. When symmetry is spontaneously broken, the system is in a state represented by one of the points of the circle. In contrast to the previous example, the symmetry that is broken is continuous (U(1) is a continuous group).

The 2D planar magnet is a system where a phase transition could occur as outlined above[1]. The system consists of a square lattice (say) with unit spin vectors at each site, capable of pointing in any direction in the plane. The interaction between adjacent spins is a scalar product $S_i \cdot S_k = S_i^x S_k^x + S_i^y S_k^y$, which is isotropic. This model is referred to as the 2D XY model. A physical system which actually orders by breaking U(1) symmetry is liquid ^{4}He. The

---

[1] In actual fact it does not. There is a phase transition in this system, but of a more complex nature, as we shall see in Sect. 3.6.

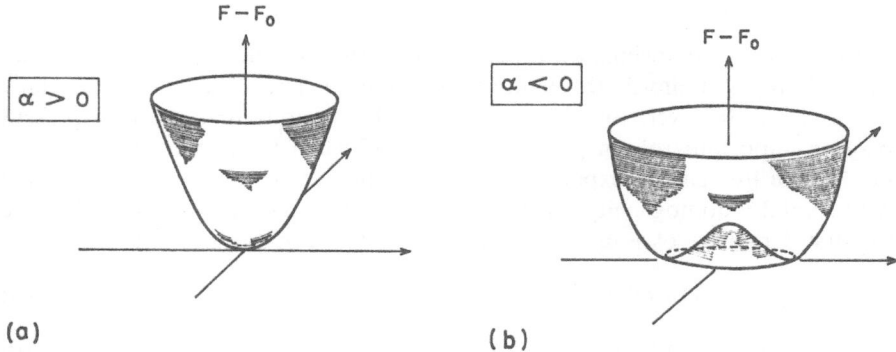

Fig. 3.3a, b. Free energy surfaces corresponding to (3.6) for $T > T_c$ and $T < T_c$. The horizontal axes represent $\mathrm{Re}\psi$ and $\mathrm{Im}\psi$ respectively

famous superfluid transition is of the above type [3.3], and the order parameter is the wavefunction of the superfluid condensate. Different realizations of the ordered state correspond to different values for the phase of ψ.

3.1.4 Order Parameter Space

The set of degenerate minima of the free energy in the ordered state constitutes the *order-parameter space*. In the first of our examples it consists of just two points, whereas in the second it is a circle (the range of variation of the phase of ψ). The order parameter space is sometimes referred to as the *manifold of internal states* [3.4]. The concept of an order-parameter space will prove very useful in the next chapter.

3.1.5 Generalized Landau Expansion

The Landau expansion of the free energy in the neighbourhood of the transition point has typically second, third and fourth-order terms in the order parameter. The third order term is needed if the phase transition is of first order [3.1].

For a scalar order parameter ψ, the expansion (3.1) is straightforward. For a general order parameter, the exercise is a bit more involved. Let the symmetry group of the free energy be the continuous group G (this is also the symmetry group of the disordered state), and let the order parameter ψ transform according to an n-dimensional representation of G. One then forms various invariants from the components ψ_α of ψ. (These are quantities which remain unchanged when the operations of the group G are applied to ψ.) Typically, these invariants are homogeneous, symmetric polynomials in $\{\psi_\alpha\}$. The invariants in the simple examples of (3.1) and (3.6) are ψ^2 and $\psi^*\psi$ respectively. A much more nontrivial case is that of superfluid ^{3}He in its diverse phases.

3.1.6 Fluctuations

The free energy expansions we have considered thus far refer to a situation where the ordering is uniform throughout the medium, i.e., ψ has the *same* value everywhere. However, the thermodynamic fluctuations that are ever-present play an important role in phase transitions [3.5]. It is therefore necessary to modify the free energy expansion so that it includes contributions associated with spatial inhomogeneity. In the simplest approximation, one would include the first derivative of ψ in an invariant form. Thus, in place of (3.1) we write

$$F = F_0 + \alpha\psi^2(r) + \beta\psi^4(r) + \tfrac{1}{2}K[\nabla\psi(r)]^2 , \tag{3.9}$$

where K is a constant. Similar modifications are made in other situations. With the gradient term included, the free energy expansion is in the so-called Ginzburg–Landau form.

We remarked earlier that when the temperature is lowered to a value below T_c, the system breaks symmetry, choosing (randomly) one out of the degenerate set of free energy minima. We can now see why the choice is random; it is essentially because of the thermal fluctuations. Referring back to Fig. 3.2, if there were no fluctuations of the order parameter, then, after cooling to a temperature below T_c, the system would continue to be in the initial state $\psi_0 = 0$, notwithstanding the fact that this state is now unstable. However, the thermal fluctuations in ψ tip the system from its unstable state into one of the stable ones. Since this occurs without the aid of any external agency, the process of ordering is an example of *spontaneous symmetry breaking*.

3.2 Conjugate Field

Consider a spin system subjected to a magnetic induction B. The magnetic ordering induced in the medium is measured by the magnetization M. The field B is *conjugate* to M in the thermodynamic sense.

In general, one can associate with every order parameter an appropriate conjugate field. Such a field may not always be realizable in the laboratory, as for example in the case of an antiferromagnet or superfluid ^{4}He. Nevertheless it has a clear physical significance. It is that field which, when applied to a disordered sample, would produce an alignment or ordering of the desired kind (e.g., staggered magnetization in the case of the antiferromagnet and phase coherence in the case of the superfluid).

The concept of the conjugate field also helps one to visualize the occurrence of spontaneous symmetry breaking. Consider a paramagnet at a temperature just above T_c. There will be large critical fluctuations leading to the (temporary) formation of large aligned domains. Such domains will lead to an internal field conjugate to the order parameter, which in turn helps drive the entire system into a state of order.

3.3 Symmetry Breaking: Further Aspects

We have seen in the previous chapter that many states of condensed matter are possible, differing in the types of translational and orientational order they have. We shall now relate this to symmetry breaking.

Let us use as an illustration the case of a liquid, which we regard as a collection of atoms arranged randomly and moving at random. If the liquid as a whole is either translated or rotated, we would not be able to recognize the difference; and this would be true for arbitrary translations and rotations. The liquid state may thus be regarded as being invariant under the transformations of the Euclidean group E(3) which is made up of rotations and translations in three dimensions. The Hamiltonian $\mathscr{H}$ of the system of atoms is invariant under the group E(3): if g is an element of this group, then

$$g^{-1}\mathscr{H}g = \mathscr{H} \ . \tag{3.10}$$

We shall use the symbol G to denote in general the symmetry group of the Hamiltonian. At sufficiently high temperatures, when all the microstates of the system become more or less equally accessible, the symmetry of $\mathscr{H}$ becomes that of the free energy F. This is the fact used in setting up the Landau theory.

When a liquid freezes to become a crystal, it breaks both continuous rotational symmetry as well as continuous translational symmetry. The symmetry group of the crystal is one of the 230 possible space groups. We shall denote the symmetry group of the broken symmetry state by H, in general. H is also called the *isotropy subgroup*. By "state" we mean of course the corresponding density matrix. At $T = 0$, naturally, broken symmetry implies that

$$h|\psi_0\rangle = |\psi_0\rangle \, (h \in H) \ , \qquad g|\psi_0\rangle \neq |\psi_0\rangle \quad \text{if} \quad g \notin H \ , \tag{3.11}$$

where $|\psi_0\rangle$ is the ground state.

We can now try to classify various states of condensed matter in terms of the symmetry that is broken. This is done in Tables 3.1 and 2. The first of these lists the group generators and the transformations associated with the symmetry groups.

Table 3.2 lists some of the important broken symmetry states. If a particular symmetry is *not* broken in the ordered state, then there is only SRO with respect to the corresponding order-parameter variable. In the case of magnetic systems, both ferromagnets and antiferromagnets break spin rotation symmetry. However, the order which results is different in the two cases. In the ferromagnet the order parameter is the magnetization, defined by $M = \langle \sum_i S_i \rangle$ where the

Table 3.1. Transformations associated with various continuous symmetries

Group	Infinitesimal generator	Finite transformations	
Translation group	k	$\exp(i k \cdot r)$:	Translation by r in the direction k
Rotation group	L	$\exp(i\theta\hat{n} \cdot L)$	Rotation in ordinary space by angle θ in the direction $\hat{n}$
Spin rotation group	S	$\exp(i\chi\hat{n} \cdot S)$:	Rotation in spin space by angle χ in the direction $\hat{n}$
Gauge group	N	$\exp(i\omega N)$:	Phase change of ω in the number field caused by the action of the number operator N

Table 3.2. Classification of condensed matter states according to symmetry breaking [3.7]. Broken symmetry is denoted by $-$, and unbroken symmetry by $+$. Numbers in parentheses indicate the number of degrees of freedom involved. A more comprehensive classification is given by *Anderson* [3.6]

Phase		Trans. symm.	Rot. symm.	Spin rot. symm.	Gauge symm.	Trans. order	Rot. order
Liquid	Normal	+	+		+	SRO	SRO
	Superfluid ^{4}He	+	+		−	SRO	SRO
Meso	Nematic	+	−		+	SRO	LRO
	Smectic A	− (1) + (2)	− (2) + (1)		+	LRO (1) SRO (2)	LRO (2) SRO (1)
	Smectic C	− (1) + (2)	−		+	LRO (1) SRO (2)	LRO
Solids	Ordinary crystal	−	−		+	LRO	LRO
	Ferro	−	−	−	+	LRO	LRO
	Antiferro	−	−	−	+	LRO	LRO

summation is over all the spins S_i. In the antiferromagnet it is the staggered magnetization defined by $N = \langle \sum_\alpha \eta_\alpha S_\alpha \rangle$ where $\eta_\alpha = 1$ for the sublattice in which the spins are "up" and $\eta_\alpha = -1$ for the sublattice in which the spins are "down".

The gauge symmetry referred to in Table 3.2 is actually the simplest of this class of symmetries. The symmetry group U(1) mentioned therein is connected with transformations on boson field operators, and is therefore relevant to liquid ^{4}He and superconductors. In both cases, the order parameter is of the form $\psi = |\psi| \exp(i\theta)$, and gauge transformations modify the phase θ. The symmetry that is broken in the formation of the superconducting or the superfluid state is gauge symmetry, and a gauge transformation applied to the ordered state changes θ to some other value corresponding to a different ordered state, recall (3.8). More intricate broken gauge symmetries are involved in systems in which appropriate "internal" degrees of freedom are associated with the ordering process. Examples of this include liquid crystalline phases and the phases of superfluid ^{3}He.

3.4 Goldstone Modes

The perfectly ordered state (ψ_0 uniform everywhere) is a state of minimum energy. States immediately above it are associated with elementary excitations involving spatial and temporal fluctuations of the order parameter. As with lattice vibrations, these elementary excitations can be characterized by a frequency-wave vector relation $\omega = \omega(k)$, when G (the unbroken symmetry group) is continuous.

An interesting question is what happens when $k \to 0$. It turns out [3.8] that if the forces are of short range, then $\lim_{k \to 0} \omega(k) = 0$, i.e., the dispersion curve continuously approaches the origin as schematically illustrated in Fig. 3.4. Such a behaviour is indeed observed for spin waves in ferromagnets and antiferromagnets. The spin waves are elementary excitations associated with the fluctuations of the order parameter.

The $k = 0$ mode, called the *Goldstone mode* [3.9], is of special significance.

This may be understood by considering a simple example. Figure 3.5a shows a (one-dimensional) line, which is obviously invariant under all continuous translations along the line. The symmetry group is thus T(1), the group of continuous translations in 1D. Figure 3.5b is a lattice which, unlike the line,

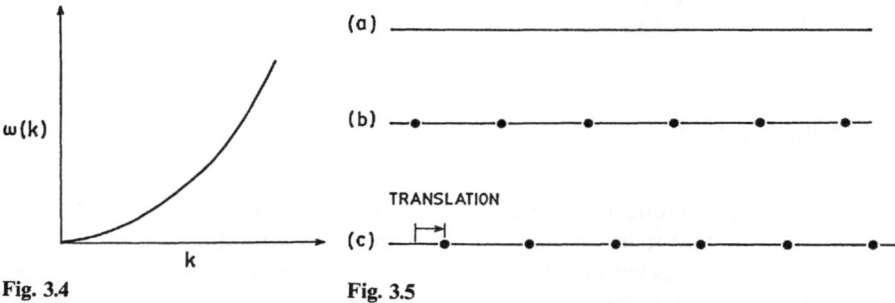

Fig. 3.4   Fig. 3.5

Fig. 3.4. Schematic plot of the dispersion curve of the (harmonic) fluctuations of the order parameter. In the case of spin waves in a ferromagnet $\omega(k)$ is parabolic in the small k limit (as in the illustration here), while for phonons it is linear

Fig. 3.5. (a) shows an infinite line. This has the symmetry $T(1)$. One may break this symmetry to obtain a lattice as in (b). The latter has the discrete symmetry $\mathbb{Z}$. The Goldstone mode essentially translates (b) taking it through lattice configurations such as that in (c). In effect, the Goldstone mode causes a continuous line to be traced and thereby restores the broken symmetry

remains invariant only under lattice translations. The lattice thus breaks the symmetry $T(1)$. If now we imagine atoms to be located at the lattice sites and allow them to vibrate under the influence of interatomic forces, we obtain the corresponding normal modes, i.e., the phonons. As is well known, the $k = 0$ mode corresponds to a *uniform* translation of the lattice along the line. Such a mode takes the lattice of Fig. 3.5b through configurations like those in Fig. 3.5c, which correspond to other possibilities for the same broken symmetry. Thus the Goldstone mode takes the system through the set of all the possible ordered states (i.e., of the various degenerate ground states of the system). The mode costs no energy (since $\omega = 0$), but since $\omega = 0$, the system stays put for all practical purposes in the particular broken symmetry state into which it condensed initially.

3.5 Generalized Rigidity

We are quite used to the idea that a solid is rigid; when we move one end of a metal rod, the other end moves through the same distance. It is interesting to note that this is connected with order present in the system!

All ordered systems (a crystal is just one example) act as a rigid entity insofar as the order parameter is concerned. If a perturbation is applied to the system at one point which disturbs the order in that neighbourhood (e.g a force which disturbs the density near one end of the rod), the system transports the disturbance over *macroscopic* distances, even though there may be no long-range forces in the system. The system acts cooperatively as a "rigid" entity in responding to the applied disturbance. This rigidity associated with the order parameter is referred to as *generalized rigidity* [3.6][2].

3.6 Quasi LRO

We return to quasi long-range order, to which a brief reference was made in the last chapter. While it is customary and convenient, in solid state physics, to introduce many ideas with the help of one- and two-dimensional models, it turns out that conventional *crystalline* order is actually not possible in these dimensions for physical reasons—it is destabilized by fluctuations [3.10, 11].

Ordering in low-dimensional systems has been, of course, the subject of intensive study for many years now, from the time of the classic work of *Onsager*

---

[2] We are referring here to a rigidity born out of long-range order. It so happens that, in a crystal, order-parameter rigidity becomes equivalent to elastic rigidity. But the converse need not hold, i.e., the existence of elastic rigidity does not necessarily imply structural LRO: example, glass.

[3.12] on the 2D Ising model. The pair of integers (d, n), where d is the physical dimensionality of the system and n is the number of components in the order parameter, are crucial in deciding the ordering behaviour of the system. Investigations in the late sixties showed that unlike the $(2, 1)$ case (the 2D Ising model), the $(2, 3)$ case (e.g., the 2D Heisenberg magnet) does not order [3.13]. The $(2, 2)$ system appeared to be a border-line case. This was confirmed by *Kosterlitz* and *Thouless* [3.14] and later by *Nelson* and *Halperin* [3.15], who showed that this system has phase transitions, but not of the usual kind.

Figure 3.6 paraphrases some of these results, in the context of 2D crystalline order and melting. Let us start with the low temperature phase. The atoms are arranged in a nearly hexagonal array. There is long-range bond orientational order, i.e.

$$\lim_{r \to \infty} G_6(r) = \text{constant} , \qquad \text{where} \tag{3.12}$$

$$G_6(r) = \langle Q_6(0) Q_6(r) \rangle , \qquad Q_6(r) = \exp[6i\theta(r)] , \tag{3.13}$$

$\theta(r)$ characterising the orientation of the set of bonds centred at r, cf. (2.5). However, the system is *not* a perfect crystal, and there is no *translational* LRO.

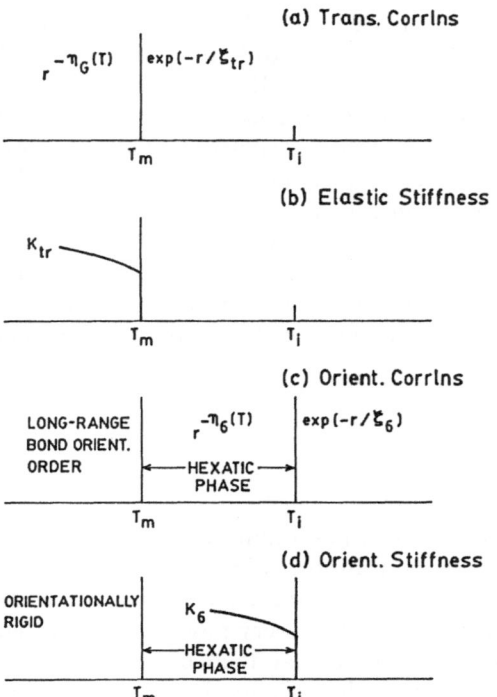

Fig. 3.6a–d. Summary of phase transitions of a 2D hexagonal array. At low temperatures one does not have perfect positional ordering, and translational correlations decay according to a power law. At T_m there is a melting leading to short range translational order. Correspondingly the elastic stiffness K_{tr} vanishes (b). Below T_m there is a perfect bond orientation, but for $T_m < T < T_i$ there is a power law decay (c). The orientational stiffness vanishes above T_i (d), leading to an isotropic liquid

One finds

$$G(r) \sim r^{-\eta} , \tag{3.14}$$

i.e., translation correlations decay rather slowly, according to a power law. The behaviour is thus intermediate between SRO (characterized by an exponential decay) and true LRO. As a result, instead of sharp Bragg peaks in the diffraction pattern, i.e., instead of

$$S(q) \sim \delta(q - G) , \tag{3.15}$$

where $S(q)$ is the structure factor and G denotes a reciprocal lattice vector, one has power-law singularities of the form

$$S(q) \sim |q - G|^{-2 + \eta_G} . \tag{3.16}$$

Property (3.14) is referred to as quasi-LRO [3.16]. Associated with it is a nonvanishing stiffness constant K_{tr}. At a certain temperature T_m, the quasi-LRO of $G(r)$ is destroyed, and K_{tr} jumps to the value zero. A new phase occurs in which $G_6(r)$ has a power law decay and $G(r)$ has an exponential decay. The system is now somewhat like a liquid crystal, and this mesophase is referred to as the *hexatic* phase.

When the temperature is raised still further to a certain value T_i, the quasi-LRO associated with orientations is destroyed, and the system becomes a normal isotropic liquid possessing SRO with respect to both orientations and translations. An important point about the transitions at T_m and T_i is that they are controlled by the breaking up of certain topological defects in the system. We shall see in the next chapter what is meant by topological defects. The transition at T_m is associated with the breaking up of dislocation pairs; that at T_i, with the breaking up of each dislocation into its constituent pair of disclinations.

The mechanism described above for 2D melting has given rise to considerable discussion [3.17]. There is very recent experimental evidence [3.18] for the existence of the intermediate hexatic phase. However, more work needs to be done to determine conclusively the nature and the role of the topological defects involved in the process of melting.

3.7 Overview

Condensed matter systems frequently exist in equilibrium states that break the symmetry implied by various groups of transformations which leave the microscopic Hamiltonian invariant. The fact that the Hamiltonian possesses some symmetry does not automatically imply that every physical state of the system also possesses the same symmetry. In particular, there could be thermodynamic states which lack the full symmetry of the Hamiltonian. The crystalline state, the liquid crystalline state, etc. are examples of (spontaneously) broken symmetry.

The Landau theory approaches the problem of symmetry breaking via a phenomenological free-energy expansion. While it provides a convenient setting to understand the essentials of symmetry breaking, at least at a pheno-menological level, it must be augmented by a careful treatment of the effect of fluctuations in different systems.

The notion of symmetry breaking actually tells us something more than just the occurrence of ordered states. Under certain circumstances, one can make statements about the nature of elementary excitations, in particular about the Goldstone modes. In fact, the symmetry G of the disordered state and the symmetry H of the ordered state decide the kinds of (topological) defects possible in the ordered system. This is the subject of the next chapter.

4. Defects and Topology

The ordered states which we considered in the preceding chapter are spatially uniform, i.e., the order parameter ψ has the same value everywhere in space. In practice, however, ψ could be a function of r on account of imperfections. Our interest here is in *topological* defects (like dislocations), that are related intrinsically to symmetry breaking. We shall therefore not be concerned with non-topological defects such as interstitials or substitutional impurities in crystalline solids. Topological defects comprise both singular and nonsingular configurations [4.1]. The latter are sometimes referred to as *textures*.

In this chapter we discuss how topological defects can be classified systematically, given the symmetry group G of the disordered state and the isotropy subgroup H of the ordered state [4.2–6]. The method works well when a continuous symmetry other than translational symmetry (e.g., rotational symmetry or gauge symmetry) is broken. Systems with broken translational symmetry present some complications which will be briefly mentioned towards the end of the chapter.

4.1 Basic Strategy

Figure 4.1 shows two singular defects possible in a planar magnet. In both examples, the order-parameter field $\psi(r)$ (here, a planar vector S of fixed magnitude) becomes singular at a certain point. A convenient way of detecting whether or not a singularity exists in a given region is to monitor $\psi(r)$ as one traverses a closed contour C, called a *Burgers circuit*, surrounding the suspected singularity. When this is done for the defect in Fig. 4.1a, for example, we find that in each *local* region like X or Y, . . . the ordering appears to be nearly perfect, but as C is traversed, ψ shows an r-dependence. The field $\psi(r)$ can be regarded as built up from the various possible states of perfect ordering. Since the order-parameter space V comprises the latter, it is natural to expect that defects can be described in terms of the properties of this space. Figure 4.2 illustrates how.

Shown on the left of Fig. 4.2 are various spin patterns. In each case one traverses a circuit C to assess the nature of the pattern. The order-parameter space V is the set of all directions of S, i.e., a circle (shown on the right). As the traverse is made, the state of local order is determined by referring to V. When a

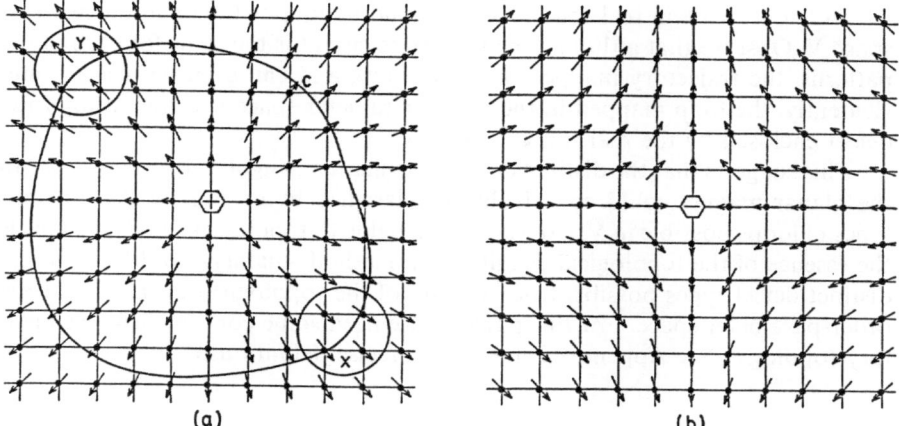

Fig. 4.1a, b. Some possible singular point defects in a planar magnet. Observe that the system appears to be perfectly ordered in small local regions like X and Y

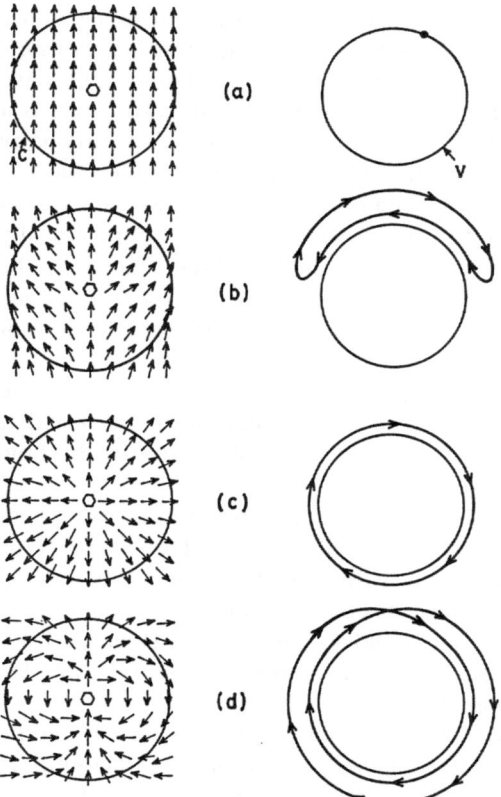

Fig. 4.2a–d. On the left are illustrated some of the possible spin configurations of the planar magnet. While **(a)** represents perfect ordering, all the others represent states with defects. On the right are sketched the loops in V which result when a round trip is made in real space. Observe that the loop corresponding to defect **(b)** can be shrunk to a point

round trip is made in real space, a corresponding loop is traced in the abstract space V. Observe that although the real space circuit is the same for each of the patterns, the trajectory mapped in V can vary, and can even be just a point. Evidently, the loop mapped in the order-parameter space is a signature of the defect enclosed by the real space contour C.

This suggests that the distinct types of elementary singular defects possible in the planar magnet could conceivably be classified by examining the different types of loops possible in V, a question amenable to topological analysis. This is the essence of the topological classification method—namely, to determine the distinct defect types possible from a study of the topological properties of the order-parameter space. For this purpose, we need some concepts from elementary topology. (See Appendix E for more precise definitions.)

4.2 Some Basic Concepts of Topology

Topology is a branch of geometry concerned with the way in which figures are "connected" and *not* with their shape or size. It deals with those geometric properties which are unchanged by topological mappings (or *topological transformations*), i.e., by mappings which are one-to-one, continuous and have continuous inverses. Intuitively, a topological transformation of an object is a continuous deformation achieved by bending, stretching, or twisting, but *not* by breaking or tearing. The term used for the geometric figure or the object is *topological space*. (A precise definition is given in Appendix E.) Two topological spaces are said to be *homeomorphic* to each other if one can be topologically transformed to the other. A triangle is homeomorphic to a circle, since it can be continuously deformed to the latter, but it is not homeomorphic to a line segment. The concept of homeomorphism in topology is analogous to that of congruence in geometry.

Consider, on the plane of the paper, a simple disc (Fig. 4.3a). Any closed curve lying entirely in it can be continuously deformed and shrunk to a point. The disc is said to be a *simply connected* topological space. By this token, a disc with a hole (Fig. 4.3b) is clearly *multiply* connected.

Consider a closed, non-self-intersecting curve C on the surface of a sphere (Fig. 4.4a). Without cutting, this curve can be continuously altered in shape as well as slid on the surface to make it coincide with another loop like C'. Curves C and C' are said to be *homotopic* to each other. Using this definition we immediately see that while the loops C_1 and C_1' on a torus (Fig. 4.4b) are homotopic, they are not so with respect to either C_2 or C_3. Further, C_2 and C_3 are not homotopic to each other, either.

It is clear that loops on a surface can be organized into distinct *homotopy classes* such that a loop in one class, while remaining homotopic to other loops in the same class, cannot be deformed to become a loop of another class. The

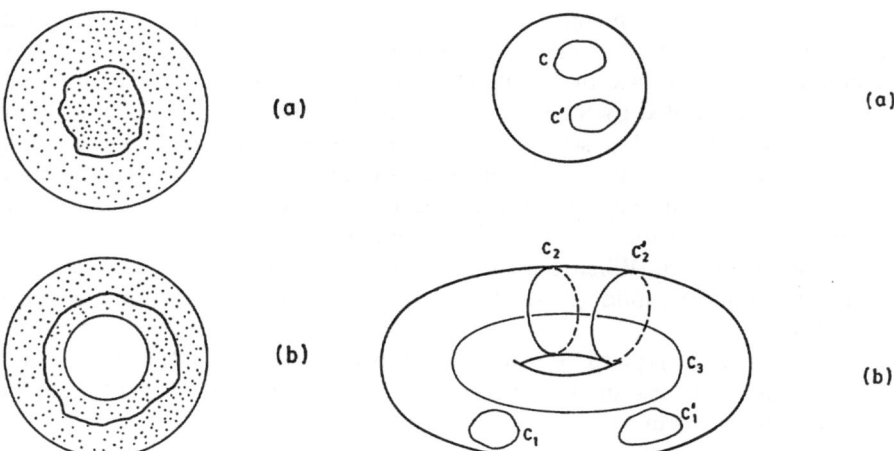

Fig. 4.3a, b. Domains with different connectivities

Fig. 4.4. Illustration of the homotopy of loops on surfaces (*see text*)

homotopic class structure of loops is intimately related to the topological properties of the space in which the loops exist.

4.3 Continuous Groups and Topological Spaces

The order-parameter space V is derived from the topological spaces associated with continuous groups of transformations. It is therefore useful to recall briefly the topological or parameter spaces of some commonly occurring continuous groups.

Associated with a continuous group of transformations are two spaces: (1) a topological space G of group parameters, and (2) a geometric space M on which the group G acts. Consider for example the group SO(2) of rotations in the two-dimensional space $M = \mathbb{R}^2$ described by the transformations

$$\begin{pmatrix} x' \\ y' \end{pmatrix} = \begin{pmatrix} \cos \theta & \sin \theta \\ -\sin \theta & \cos \theta \end{pmatrix} \begin{pmatrix} x \\ y \end{pmatrix} , \tag{4.1}$$

or, symbolically, $r' = T_\theta r$. Here the rotations $\{T_\theta\}$ parametrized by a single parameter θ constitute the group elements, and r, r' belong to the space M. Two rotations combine in the space G according to the relationships

$$T_\theta T_{\theta'} = T_{\theta + \theta'} ,$$

$$I = T_0 \text{ (identity)} , \tag{4.2}$$

$$T_\theta^{-1} = T_{-\theta} \text{ (inverse)} .$$

Since $T_\theta = T_{\theta+2\pi l}$ (l = integer), the points θ and $\theta+2\pi l$ have to be identified. Thus the group space G can be taken, conveniently, to be a circle of unit radius, denoted S^1. (The surface of a unit sphere in an $(n+1)$-dimensional Euclidean space E^{n+1} is denoted by S^n. S^0 is the "sphere" in 1D, and consists of just two points—the ends of a line segment. S^2 is the familiar surface of a sphere in E^3.)

The continuous group SO(3) of proper rotations in 3D can be parametrized in various ways. One way would be to specify every rotation by a vector whose direction represents the direction of rotation, and whose magnitude represents the angle of rotation. When drawn from a common origin, the tips of all such vectors lie in or on a solid sphere of radius π. This sphere is the parameter space of SO(3)[1].

The group SU(2) is well known in the study of electron spin. It is the group of 2×2 unitary matrices of unit determinant. Every element of SU(2) can be expressed in the form

$$Q = a_0 \underline{I} + \mathrm{i}(a_1 \underline{\sigma}_1 + a_2 \underline{\sigma}_2 + a_3 \underline{\sigma}_3) \ , \tag{4.3}$$

where $\underline{I}$ is the (2×2) unit matrix, $\underline{\sigma}_1, \underline{\sigma}_2, \underline{\sigma}_3$ are the well-known Pauli matrices and the real parameters a_i satisfy the relation

$$a_0^2 + a_1^2 + a_2^2 + a_3^2 = 1 \ . \tag{4.4}$$

The condition represents S^3, so that the parameter space of SU(2) is S^3. S^3 is a simply connected space. The group SU(2) is called the universal covering group of SO(3). There are two elements of SU(2) (differing by a sign) corresponding to each element of SO(3).

4.4 The First or the Fundamental Homotopy Group and Defects

We return to a consideration of the topological properties of the order-parameter space V. Our interest in this sprang from the observation that a traversal around a singular defect in real space produces in V a mapping which can be used to characterize the defect.

4.4.1 Burgers Circuit

The circuit C in Fig. 4.2 is a typical example of the Burgers circuit that "*encloses*" a defect. In general the circuit is an appropriate sphere S^r. The number r depends

---

[1] Note that no two points inside this sphere represent the same rotation, whereas each pair of antipodal points on its surface corresponds to the same rotation (through an angle π) about two oppositely directed axes. This important fact makes the parameter space of SO(3) doubly connected.

both on the dimensionality d of the ordered medium and the dimensionality d' of the defect: One has [4.3, 7]

$$r = d - d' - 1 \; . \tag{4.5}$$

Thus, to surround an isolated point defect ($d' = 0$) in a three-dimensional medium ($d = 3$), we need the Burgers sphere S^2, whereas for a similar defect in a two-dimensional medium we need the circuit S^1. A traversal over S^r is mapped into a corresponding image in V, and we must study the homotopic properties of these maps. We will be mostly concerned with Burgers circuits topologically equivalent to S^1.

4.4.2 Closure Misfit

We have indicated earlier that the purpose of traversing the Burgers circuit is to obtain a measure which, when compared with the corresponding quantity for the perfect (reference) medium, would indicate whether or not a defect is enclosed. To sharpen our ideas, let us recall the familiar example of the dislocation. Figure 4.5 shows a dislocation in a square lattice. The Burgers circuit in the imperfect lattice on the left has an image in the perfect lattice shown on the right. The latter is no longer a closed curve.

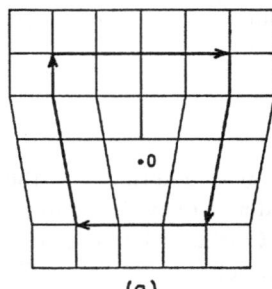   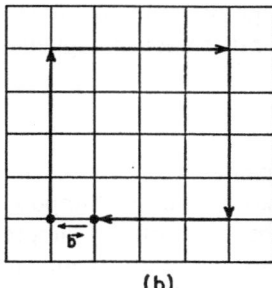

(a)   (b)

Fig. 4.5a, b. Dislocation in a square lattice

The *closure misfit* or the *closure failure* b is a measure of the defect at 0. In a continuum description, b is given by

$$b = \oint du(r) = \oint dl \, \partial u/\partial l \; , \tag{4.6}$$

i.e., the line integral of the elastic displacement field $u(r)$.

In the case of the planar magnet, the closure failure does not manifest itself as a spatial gap. Since magnetization occurs by the breaking of spin orientation symmetry, the quantity we monitor is the angle θ which the spin vector makes with a reference direction, say the x-axis. When we move around C in Fig. 4.1a, we find that the vector we started with has rotated through 2π. As in the previous example, the closer failure ϕ is in general given by the line integral over the

Burgers loop C, of the field $\theta(r)$:

$$\phi = \oint_C d\theta(r) = \oint_C dl \; \partial\theta/\partial l \; . \tag{4.7}$$

The corresponding rotation in the perfect ferromagnet would of course be zero.

In general, the nature of the misfit depends on the nature of the order parameter. Only in the special case where translational symmetry is broken does it manifest itself as a gap in space.

4.4.3 Fundamental Group

Consider now a general topological space V and let x be a point in this space. Draw all possible loops starting and ending at x. We already know that these loops can be organized into distinct homotopic classes which we denote by $[e]$, $[a]$, $[b]$, . . . , where $[e]$ is the class of loops which can be collapsed to a point at x. It turns out (see Appendix E) that the classes $[e]$, $[a]$, $[b]$ . . . can be assigned the status of group elements; and that the collection $[e]$, $[a]$, $[b]$, . . . forms a group with $[e]$ as the identity element. This group, denoted by $\Pi_1(V, x)$, is called the *fundamental group* at x or the *first homotopy group* based at x.

$\Pi_1(V, x)$ characterizes the homotopy properties of loops in V at x. As the choice of x is arbitrary, one is actually interested in a group $\Pi_1(V)$ that characterizes the entire space V. Such an object is realized by considering loops in V that are not tied to a point x, but are allowed to slide freely, i.e., are "freely homotopic". This is elaborated upon in Appendix E.

4.4.4 $\Pi_1(V)$ and Defects

Consider a system the symmetry of whose Hamiltonian is described by the continuous group of transformations G. Suppose the system can exist also in an ordered state. Let H denote the isotropy subgroup of G (the ordered state is invariant only under transformations belonging to H). The left cosets $\{gH \mid g \in G\}$ of H in G form a space, denoted by G/H (see [4.4]). It can be shown that this space is in one-to-one correspondence with the order-parameter space V (see Appendix E). The first homotopy group $\Pi_1(G/H)$ should then tell us something about singular topological defects in the system which can be detected with Burgers circuits that are loops. More specifically, corresponding to every conjugacy *class* of the group $\Pi_1(G/H)$, there is a distinct topological defect. If, in particular, $\Pi_1(V)$ is Abelian, each element of Π_1 corresponds to a distinct defect.

The basic steps involved in the topological classification method are as follows:

  i) Identify G, the symmetry of the parent (disordered) state.
 ii) Identify H, the isotropy subgroup of G which describes the surviving symmetry of the broken-symmetry phase.

iii) Determine the order-parameter space $V = (G/H)$.
iv) Find $\Pi_1(V)$.
v) Identify the conjugacy classes of $\Pi_1(V)$.

There are as many distinct defect types as there are classes in the first homotopy group $\Pi_1(V)$.

4.5 Some Examples

We consider a few examples to illustrate how the topological classification method works. Consider first the planar magnet. Here $G = SO(2)$, while $H = \{e\}$, the trivial group. Therefore $V = G/H = G$, and $\Pi_1(V) = \Pi_1[SO(2)]$. We have seen earlier that the topological space of $SO(2)$ is S^1. Therefore $\Pi_1(V) = \Pi_1(S^1)$. The question is now: How many homotopy classes of loops can we draw in S^1? While there is a formal way of obtaining the answer, one can intuitively guess from the example in Fig. 4.2 that there is a countable infinity of classes. The loops in each class wind around S^1 an integer number of times in a clockwise or an anticlockwise direction. The homotopy classes of loops in S^1 can thus be labelled by the elements of the integer group $\mathbb{Z}$ with elements $\{\ldots -2, -1, 0, 1, 2, \ldots\}$, 0 as the identity, and addition as the law of combination. One finds

$$\Pi_1[SO(2)] = \Pi_1(S^1) = \mathbb{Z} \ . \tag{4.8}$$

The topologically distinct, singular, point defects of the planar magnet are in one-to-one correspondence with the classes of the group $\mathbb{Z}$. Since the group is abelian, the correspondence is with the elements of $\mathbb{Z}$ themselves. Each defect is thus labelled by an integer. (This is just the *winding number* of the vector field of the order parameter in the present case.) Field patterns corresponding to the same winding number can be transformed into each other. Further, patterns corresponding to defects with different winding numbers can be superposed, leading to a new pattern characterized by a winding number which is the algebraic sum of those of the patterns superposed. Winding numbers may be regarded as quantum numbers of topological origin.

Our next example is superfluid ^{4}He. From the preceding chapter we know that $G = U(1)$, while $H = \{e\}$, the trivial group. Therefore

$$\Pi_1(V) = \Pi_1(G/H) = \Pi_1[U(1)] = \Pi_1(S^1) = \mathbb{Z} \ . \tag{4.9}$$

Thus the *line* defects in bulk He-II are vortices with integer winding numbers.

As our last illustration, we consider the three-dimensional classical Heisenberg model. The order parameter is a vector of constant magnitude, so that the order-parameter space is S^2. The line defects are given by the classes of $\Pi_1(S^2)$. Referring back to Fig. 4.4a we see that all the loops which can be drawn on S^2 are homotopic to each other, and further can be shrunk continuously to a point. That is,

$$\Pi_1(S^2) = \{e\} \ . \tag{4.10}$$

Thus there are no topologically stable line defects in this system.

All the examples above are elementary. In every case, information about the defects could have been obtained without resorting to homotopy theory. However, when the order parameter is more complicated, as in the case of liquid crystals or superfluid ^{3}He, the homotopy technique comes into its own.

4.6 Stability

There are two types of stability we must consider: topological stability and physical stability. Figure 4.6 offers a simple illustration. We have here a (hypothetical) 1D medium with two possible states of ordering (i.e., free energy minima). As $x \to \pm\infty$, one of these two states must be approached, or else the gradient terms in the free energy would lead to an infinite total energy. Suppose one minimum is attained as $x \to -\infty$ while the other is selected as $x \to +\infty$. Then clearly, somewhere in between $\psi(x)$ has to switch between these two asymptotic values as x increases from $-\infty$ to $+\infty$. A topological soliton or a kink (of some finite width) thus occurs. While the location of the kink can be moved about, the kink itself cannot be removed. To wipe out the defect, either the infinitely long segment ABC must be flipped to the state ψ_1, or the equally long segment BCD must be flipped to ψ_2. Both processes require an infinite amount of energy. It is this energy barrier which bestows topological stability on the kink configuration.

Consider next the defect patterns of the 2D magnet illustrated in Fig. 4.2. To eliminate the defect in Fig. 4.2c, for example, all the spins in the infinite medium

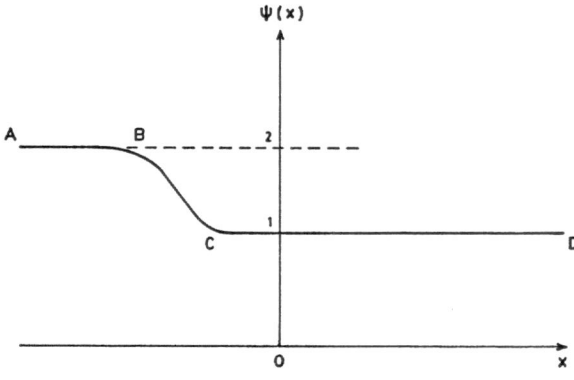

Fig. 4.6. Topological defect (or "kink") in a one-dimensional continuum. It arises because the system is clamped at two different ordered states (free energy minima) at the two end points

must be made to point in the same direction, a process which clearly will cost an infinite amount of energy. This defect is therefore topologically stable against conversion to another homotopic class (including a defect with zero winding number which would correspond to no defects). On the other hand, only a limited number of spins need to be turned around in the pattern of Fig. 4.2b to make it look like that in Fig. 4.2a. Stated differently, this defect can be eliminated by *local surgery* [4.4] alone, and is therefore topologically unstable.

Physical stability, on the other hand, refers to a minimum in the free energy. Thus while the defect in Fig. 4.2b is topologically unstable, there might not be enough energy in the system for local surgery to be effected spontaneously, i.e., by fluctuations. The defect may thus remain "frozen in" for a long time, and we would have a case of topological instability but physical stability (or at least metastability). A topologically *stable* defect, however, is generally physically stable- or at least metastable-, being separated from other topologically inequivalent configurations by energy barriers that diverge at least as rapidly as N in the thermodynamic limit $N \to \infty$.

4.7 Combination of Defects

So far we have considered only a single defect in isolation. When many defects are present, there can be defect–defect interactions leading to defect annihilation, coalescence, etc. What has homotopy to say about these possibilities? The answer depends on whether Π_1 is an Abelian group or not. If Π_1 is Abelian, then two defects A and B (corresponding to the elements a and b of Π_1) can combine to form a defect C that corresponds to the element $c = ab$ (see Fig. 4.7). However, if Π_1 is non-Abelian, complications can arise, which we do not consider here.

One peculiar situation that occurs when $\Pi_1(V)$ is non-Abelian is worth noting, especially as it is relevant to the structure of glass (Chap. 6). Consider the two line defects in Fig. 4.8. The question arises as to whether they can cross if

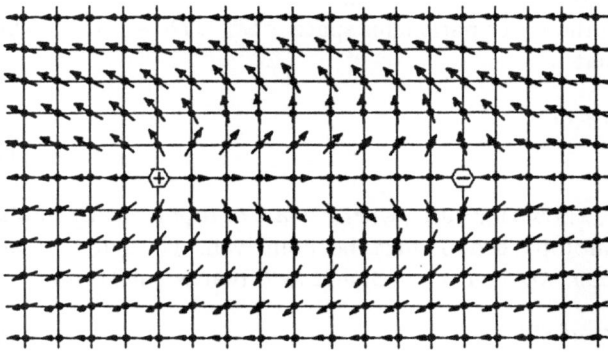

Fig. 4.7. Vortex–antivortex pair in the planar magnet system, obtained by a coalescence of the two defects in Fig. 4.1

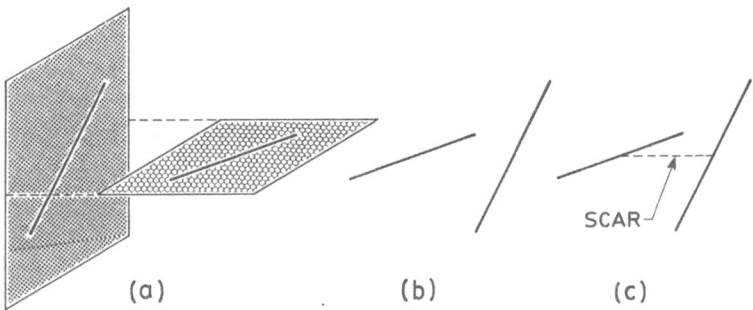

Fig. 4.8. (**a**) shows two line defects. In (**b**) the two have crossed without a scar, which happens if Π_1 is Abelian. If Π_1 is non-Abelian, the situation is as in (**c**)

they move towards each other. The answer depends on whether or not Π_1 is Abelian. If it is, they can cross freely; if not, they will entangle to produce a "scar" as in the figure.

4.8 Other Homotopy Groups

We have seen in Sect. 4.4 that the Burgers circuit in the general case is S^r, where r is given by (4.3). The rth homotopy group $\Pi_r(V)$ classifies the maps of S^r into the order-parameter space V. Table 4.1 lists some of the homotopy groups involved. For completeness, it also includes the homotopy groups relevant for textures.

We have also $\Pi_0(V)$, which is, roughly speaking, the set of disjoint pieces of V. But if V is the topological space of a continuous group, then $\Pi_0(V)$ can be given a group structure. From a physical point of view, Π_0 helps classify singular wall defects in 3D and line defects in 2D. The corresponding Burgers circuit is S^0, i.e., two points, one on either side of the singularity.

Toulouse and *Kléman* [4.7] have shown that if the order parameter is a n-component vector, then the order-parameter space is S^{n-1}. Now it is known from homotopy theory that

$$\Pi_r(S^m) = \{e\} , \quad \text{for} \quad r < m ,$$
$$\Pi_m(S^m) = \mathbb{Z} , \tag{4.11}$$

while no *general* formula is known for $\pi_r(S^m)$ when $r > m$. For topologically stable defects to occur the homotopy group must be nontrivial. Thus, for an n-component order parameter, we see from (4.3) and (4.9) that we must have

$$d - d' - 1 \geq n - 1 , \quad \text{or} \quad d' \leq d - n \tag{4.12}$$

Table 4.1. Homotopy groups Π_r needed for various defects. Besides singular defects, the requirements for bulk textures in 3D are also included

Singular defects[a]			
Defect	1D	2D	3D
Point ($d' = 0$)	Π_0	Π_1	Π_2
Line ($d' = 1$)	—	Π_0	Π_1
Wall ($d' = 2$)	—	—	Π_0
Bulk textures in 3D			
Point	Π_3		
Line	Π_2		
Wall	Π_1		

[a] A characteristic feature of singular defects is that the order parameter $\psi(r)$ varies all along the Burgers circuit. In a texture, on the other hand, it is constant. If S^r is the Burgers circuit for a singular defect of dimensionality d', that for a texture of the same dimensionality is S^{r+1}. See [4.5] for details

for nontrivial topological defects of dimensionality d'. Hence no topologically stable defects can occur for $n > d$ in such systems.

4.9   Ordered Media with Broken Translational Symmetry

The topological classification method poses problems if the continuous symmetry that is broken happens to be translational symmetry. To appreciate the problem, consider a crystal containing topological defects like dislocations. Our reference states are perfect crystals in various orientations (Fig. 4.9), and our objective is to describe the spatial pattern of the defective crystal using the reference "templates". Since the reference states are related to each other by rigid rotations, we might as well retain just one reference template and examine what rigid-body operations must be performed on it to bring it into coincidence with the local structure of the defective medium at various space points r.

Now, the set of all rigid-body operations of the reference frame forms a group G containing translations as well as rotations. G is the proper part of the full Euclidean group E(3) (i.e., G does not include inversion and reflection). Let H be the subgroup of G which leaves the reference structure invariant, i.e., H is the proper part of the conventional space group of the crystal. One could take the order-parameter space V to be the coset space G/H with G and H defined as

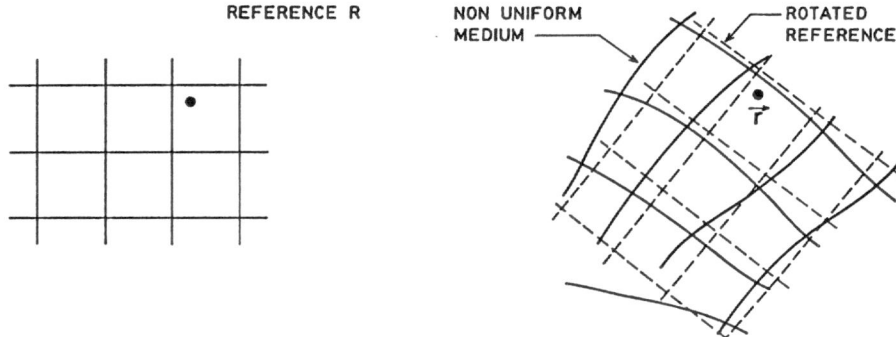

Fig. 4.9. The order parameter field in a defective crystal could be described by considering a perfect crystal *R* which serves as a reference, and considering what operations must be performed on *R* to bring it into coincidence with a local region at *r*. Different regions would require different operations. In general, deformations would be needed in addition to rigid-body motions

above, and proceed as usual. This procedure, referred to as *naive generalization*, has been employed, but the results must be treated with caution [4.4].

One problem with naive generalization is whether the reference frame can be adjusted uniquely to the distorted structure using the elements of G alone. *Mermin* [4.4] suggested that it may be necessary to extend G to include also an appropriate set of infinitesimal compressions, dilations and shears to bring the reference frame into unambiguous and precise coincidence with the local structure of the nonuniform medium. This suggestion has yet to be pursued to its conclusion (see also related remarks by *Trebin* [4.5]).

4.10 Summary

This chapter is an elementary discussion of how defects in ordered media could be topological in nature. The explicit linkage between defects and the topological properties of the order-parameter space is via the homotopy groups of the latter. The order-parameter space is itself related to the underlying symmetry of the ordered phase via the groups G and H. There is thus a connection between the broken symmetry of the ordered phase and the topological defects possible in it. The main advantages of establishing such a linkage are the possibility of a systematic classification of the defects, as well as the relating of physical processes involving defects to algebraic operations associated with homotopy groups, providing thereby deeper insight into the process concerned. Thus, defect coalescence is linked to the group product, the relative motion of defects to group action, and so on.

Noncrystalline states of matter abound in topological defects. Indeed, it is believed that some states like glass and cholesterics (blue phase) cannot occur without a very high density of such defects. Important inputs from topological analysis become necessary when one considers the many-defect problem [4.8]. A familiarity with at least the basic ideas of the homotopic classification of defects is thus very useful for a better understanding of the noncrystalline state.

5. Structures by Projection

In Chap. 2 we have seen that many structural arrangements of atoms and molecules are possible, corresponding to the various phases of condensed matter. A common feature is that the average density is generally constant in each case: equilibrium thermodynamic phases are almost always spatially homogeneous. From a purely *geometric* point of view, therefore, different phases represent the outcomes of different space-filling exercises, the building blocks being atoms, molecules or even clusters of atoms and molecules.

Of the various space-filling patterns one can generate, the crystalline one is special for several reasons. First, given a unit cell, there is only one (crystalline) arrangement one can generate. Second, the (perfect) structure has an internal rigidity in the sense that one part cannot move with respect to another, generating variations of the structure. Noncrystalline structures, on the other hand, often have "internal coordinates" such that by assigning different values to these, one can generate variations of the structure. Sometimes thermal fluctuations suffice to cause transitions involving such internal degrees of freedom. There are also situations where such degrees of freedom are frozen; different copies of the structure with different (frozen) values for the internal coordinates represent configurational degeneracy.

Penrose tilings provide an example of configurational degeneracy. As we shall see in the next chapter, such tilings can be obtained by superposing five density waves to obtain the density $\varrho(r)$ according to

$$\varrho(r) = \varrho_0 \sum_{i=1}^{5} \cos\left[k_0(e_i \cdot r) + \alpha_i\right] , \qquad (5.1)$$

where the $\{e_i\}$ are unit vectors in the plane such that $e_i \cdot e_{i+1} = \cos(2\pi/5)$ and α_i are phase angles. ($e_6 = e_1$.) Each tiling is characterized by a given set $\{\alpha_i\}$. Tilings which differ in the phase angles belong to the same local isomorphism class (see Sect. 2.2.3). The internal coordinates offer a means of describing the configurational degeneracy resulting from the variation of the set $\{\alpha_i\}$.

Configurational degeneracy is quite common, and may be expected in general whenever there is *frustration*. The concept of frustration was first introduced by *Toulouse* [5.1] and *Anderson* [5.2] in the context of spin glasses. It refers to tendencies (like attractive forces, steric hindrance, etc.) which compete in defining the stable state of the material, and in the process inhibit complete order at the molecular level. Sometimes, this reduces to a geometric problem, i.e., the inability to tile space regularly with a given basic unit. In such cases, there is usually no unique structure; sometimes there is *also* disorder (spatial randomness).

A crystal is easy to describe. There is a unit cell, group action on which generates the entire space-filling pattern. Unfortunately, there is no such unique generative prescription for noncrystalline structures. However, in recent times various methods have been evolved to derive some noncrystalline structures from suitable ordered ones via *projection* techniques. In the case of amorphous structures, *Kléman* and *Sadoc* [5.3, 4] have advanced the view that such structures could be regarded as the projections of ordered patterns in 3D spherical space (S^3) and 3D hyperbolic space (H^3). As regards incommensurate crystals and quasicrystals, it is now widely recognized that they can be regarded as projections of certain higher-dimensional crystalline structures, i.e., structures in E^n ($n > 3$) [5.5–10].

In this chapter we shall discuss these projection schemes. It would seem that the projection technique is particularly useful when there is configurational degeneracy.

5.1 Concerning Tilings

A *tiling* or a tessellation is a space-filling pattern. Let us consider first the regular tiling of the 2D spaces of constant curvature, i.e., the Euclidean plane E^2, the spherical surface S^2 and the hyperbolic surface H^2. The basic tile (prototile) is a regular polygon in the space concerned.

A polygon is regular if it is both equilateral and equiangular. In the so-called *Schläfli notation*, we denote a regular polygon by the symbol $\{p\}$ where p is the number of sides. To tessellate a 2D space we start with a regular polygon and try to put others adjacent to it so that q of them fit around any vertex without any gap or overlap. If this operation can be endlessly repeated so as to fill the whole of the space, then we have a *regular tessellation* $\{p, q\}$.

Figure 5.1 shows some examples. In the case of tilings on S^2, if we imagine S^2 to be embedded in E^3 and replace the great-circle arcs by chords, one obtains a regular polyhedron in E^3. These too can be labelled using the Schläfli notation. The only regular polyhedra possible are the Platonic solids: namely, the tetrahedron $\{3, 3\}$, the octahedron $\{3, 4\}$, the cube $\{4, 3\}$, the icosahedron $\{3, 5\}$, and the dodecahedron $\{5, 3\}$ (Fig. 5.2). No confusion should arise from the use of the same notation, $\{p, q\}$, for both the tessellation (here, of S^2) and the corresponding regular solid in a higher dimension (here, E^3).

Tessellations of the 3D spaces E^3, S^3 and H^3 are carried out in a similar manner, using $\{p, q\}$ as building blocks [5.11, 12]. Such tessellations are produced by arranging r regular blocks $\{p, q\}$ such that they share an edge, and

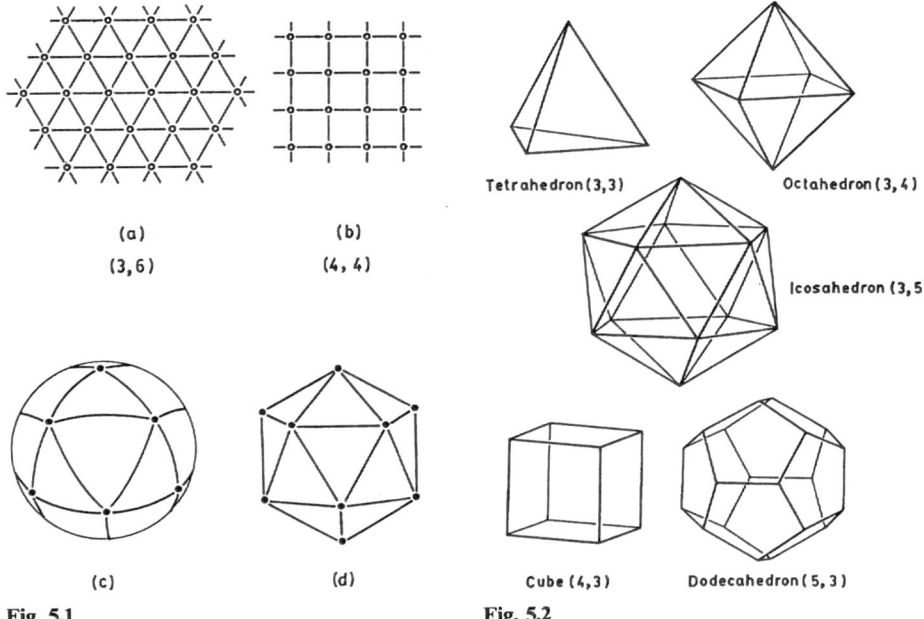

(a)
(3,6)

(b)
(4,4)

(c)

(d)

Tetrahedron(3,3)

Octahedron(3,4)

Icosahedron(3,5)

Cube(4,3)

Dodecahedron(5,3)

Fig. 5.1   **Fig. 5.2**

Fig. 5.1. Some examples of regular tessellations on E^2 [(a) and (b)] and on S^2 (c). By replacing the circular arcs in (c) with chords, one obtains a regular polyhedron as in (d)

Fig. 5.2. The five Platonic solids in E^3

the result is labelled $\{p, q, r\}$. In E^3 there is only one regular tessellation possible: namely,

$$\{4, 3, 4\} \tag{5.2}$$

which is a stacking of cubes. The possible tessellations of S^3 are:

$$\{3, 3, 3\}, \quad \{3, 3, 4\}, \quad \{3, 3, 5\}, \quad \{4, 3, 3\}, \quad \{3, 4, 3\} \quad \text{and} \quad \{5, 3, 3\} . \tag{5.3}$$

Similarly, there are eight distinct regular tessellations of H^3 [5.11], but we shall not consider them here.

As in the case of S^2, one may regard S^3 as being embedded in E^4, and replace geodesics associated with the tessellations (of S^3) by straight chords. One then obtains the 4D version of a regular Euclidean polyhedron. This object is called a *regular polytope*. A polytope is thus the next in the sequence

point,   line,   polygon,   polyhedron, ...

As in the case of regular polyhedra, the same Schläfli symbol $\{p, q, r\}$ is used for both a regular tessellation (of S^3) and the corresponding regular polytope (in E^4). In what follows, we shall occasionally use the term 'polytope' for spherical tessellation, for brevity. See also Appendices B and C for additional remarks.

5.2 Regular Polytopes

Table 5.1 is a summary of the characteristics of the regular polytopes. Of these, the tessellations of S^3 corresponding to $\{3, 3, 5\}$ and $\{5, 3, 3\}$ are the ones most favoured for generating amorphous structures in E^3. The tessellation $\{3, 3, 5\}$ is a tiling made up of 600 regular (spherical) tetrahedra, five of them sharing each edge. There are 120 vertices, each with an ideal icosahedral coordination. Being rich in tetrahedra, this polytope is convenient for generating DRP structures (in E^3). The other important polytope $\{5, 3, 3\}$ is made up of 120 regular dodecahedra, three sharing every edge. This polytope is useful for generating CRN structures in E^3.

Table 5.1. Summary of the characteristics of regular polytopes in E^4. Here V denotes the number of vertices, E the number of edges, F the number of faces and P the number of polyhedra. We note that $V + F = E + P$. This is a generalization of the familiar Euler relation $V + F = E + 2$ in three dimensions. Polytopes $\{p, q, r\}$ and $\{r, q, p\}$ are dual to each other [5.11]. This is reflected in the interchanges $V \leftrightarrow P$ and $E \leftrightarrow F$ for such pairs

Name	Nature of P	Schläfi symbol	V	E	F	P
Regular simplex	Tetrahedron	$\{3, 3, 3\}$	5	10	10	5
16-cell	Tetrahedron	$\{3, 3, 4\}$	8	24	32	16
600-cell	Tetrahedron	$\{3, 3, 5\}$	120	720	1200	600
24-cell	Octahedron	$\{3, 4, 3\}$	24	96	96	24
Hypercube	Cube	$\{4, 3, 3\}$	16	32	24	8
120-cell	Dodecahedron	$\{5, 3, 3\}$	600	1200	720	120

5.3 Amorphous Structures from Mappings of Polytopes

As already noted, a characteristic feature of an amorphous material is its lack of LRO. Nevertheless, it is pertinent to ask whether amorphous structures are in any way related to ordered ones. The question is not without meaning—since a glass can devitrify, for instance. Also, near-neighbour coordinations in an amorphous solid and its crystalline counterpart are often very similar. This has prompted many investigators to seek geometric relationships between an amorphous structure and the corresponding crystalline one.

Kléman and *Sadoc* [5.3] have also sought a similar relationship, but with a difference. It will be recalled (Chap. 2) that Bernal found DRP structures to be dominated by tetrahedra. If we now attempt to compare the DRP with a tiling of E^3 by regular tetrahedra, then we face a problem right away, because a tessellation of E^3 by $\{3, 3\}$ is not possible; recall (5.2). Kléman and Sadoc suggested that the DRP should be related to the tessellation $\{3, 3, 5\}$ of the *spherical* space S^3 by $\{3, 3\}$.

Both the tessellation $\{3, 3, 5\}$ and the DRP share the same local order or local topology. In $\{3, 3, 5\}$, every bond is shared by five (regular) tetrahedra. In the DRP, too, most of the bonds are shared by five (slightly distorted) tetrahedra. One important difference between $\{3, 3, 5\}$ and the DRP is that in the latter there are also bonds shared by 4, 6 or even 7 tetrahedra, i.e., the local neighbourhood around these bonds is different. Kléman and Sadoc argue that such departures arise during the mapping of the tiling on S^3 onto E^3, and that such departures are in fact responsible for the "amorphization" of the initial regular structure.

The problem of building an amorphous structure can now be sharpened and reduced to two essential steps: (i) Selecting an initial ordered structure (in curved space), and (ii) defining a procedure for mapping or projecting this ordered structure onto flat space. We have already discussed (although briefly) ordered structures in curved space. It remains now to discuss the projection scheme. However, before doing so, a few remarks are necessary on line defects in amorphous structures.

5.4 Line Defects in Amorphous Structures

The notion of a line defect in an amorphous structure might seem strange, but it ceases to be so if *local* order is considered. Historically, lines of "anomalous" coordination were first introduced by *Frank* and *Kasper* [5.13] in their studies on the structure of complex crystalline alloys. Earlier, *Frank* [5.14] had pointed out that, when required to close-pack, atoms interacting with each other via central forces prefer (for minimisation of the energy) icosahedral rather than a face-centred-cubic or hexagonal-close-packed coordination. It is natural, then, to regard the formation of complex structures as being dictated primarily by local aggregations; and to suppose that in such structures, every atom has an icosahedral environment, implying a coordination number $Z = 12$. But a problem arises because, as with regular tetrahedra, one cannot have a space-filling structure (in E^3) in which *all* atoms have icosahedral coordination. Frank and Kasper then argued that if coordinations $Z = 10$, 14, 15 and 16 were also permitted, the shortfalls in icosahedral packing could be made up. They also showed (using topological arguments) that atoms with anomalous coordination (i.e., $Z \neq 12$) formed infinite chains, with Z remaining constant along each chain (taking values such as 10 or 14, . . .). A Frank–Kasper crystal is therefore an ordered structure in which some atoms have icosahedral coordination while others with $Z \neq 12$ form *Frank–Kasper chains*. Further, the chains are themselves spatially ordered.

Figure 5.3 shows how β-tungsten may be viewed in this picture. There are 8 atoms in the unit cell, two of them with $Z = 12$, and the remaining with $Z = 14$. Atoms with $Z = 14$ form Frank–Kasper chains which are organized into three regular, non-intersecting arrays.

The Frank–Kasper line shows how one could identify "lines of anomalies" using local topology as the guideline. Frank–Kasper lines exist in DRP structures. To see this, let us consider first Fig. 5.4, which compares the coordination cages for $Z = 12$ and $Z = 14$. Each cage consists of two biprisms. In the biprism making up the icosahedron, we find that the bond AB is shared by five tetrahedra; whereas the corresponding bond $A'B'$ in the biprism of the $Z = 14$ cage is shared by six tetrahedra. In fact, the latter biprism is obtained by introducing a wedge into the former. A Frank–Kasper line can thus be viewed

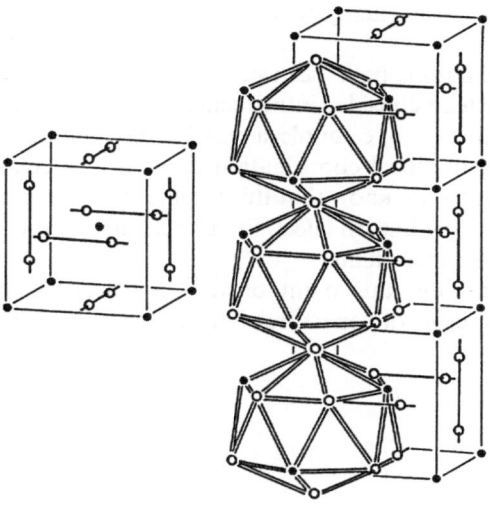

◀ Fig. 5.3. On the left is a unit cell of β-tungsten. On the right, three $Z = 14$ cages are illustrated. Their axes link up to form a Frank–Kasper chain

Fig. 5.4a, b. On the left are the polyhedral cages for $Z = 12$ and $Z = 14$, the dotted circles representing the atoms at the centre. On the right are shown the prisms of which the cages are made. Bonds shared by 5 or 6 tetrahedra form the backbones of the cages, and the latter link up to form Frank–Kasper chains

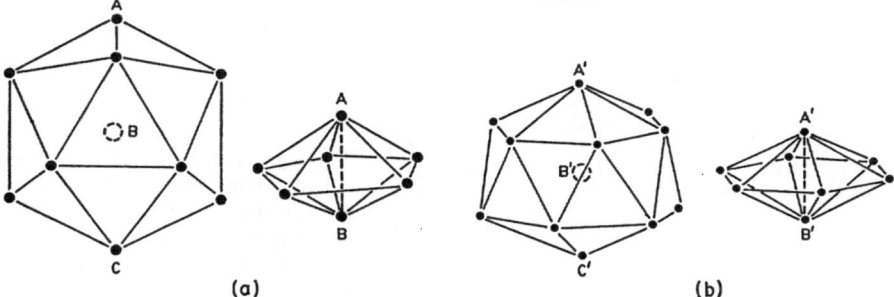

(a)   (b)

also as a continuous chain of anomalous bonds, i.e. bonds shared by either more than, or less than, five tetrahedra. Of course, this number remains constant along a given chain, since Z does *not* vary along such a chain.

The DRP is a *random* structure, and it is built up from tetrahedra. Therefore some of the bonds in it must necessarily be shared by an anomalous number (4, 6, 7. . .) of tetrahedra. And, for the same topological reasons as invoked by *Frank* and *Kasper* [5.13], such bonds join up to form (Frank–Kasper) lines. These we identify as topological defects, having used local topology to characterize them.

5.5 Disclinations and Frank–Kasper Chains

A disclination is a rotational analogue of a dislocation. Figure 5.5 shows examples in which a disclination is created essentially by either removing or adding a wedge. Disclinations are generally not found in crystals because of their cost in energy. However they may exist—in energy-reducing arrays or tangles—in amorphous structures.

In the example of Fig. 5.5, disclinations have been introduced into regular Euclidean tessellations. One may instead consider introducing a wedge disclination into a tessellation in spherical space. For example, a possible disclination in $\{3, 3, 5\}$ essentially involves altering the local coordination along a particular line from $Z = 12$ to $Z = 14$. From Fig. 5.4 we know that this implies that a wedge is inserted in every icosahedron along the line. In effect, therefore, a disclination in $\{3, 3, 5\}$ is nothing but a Frank–Kasper chain.

Once again we emphasize that, from this point of view a defect in an amorphous structure is identified in terms of departures in local topology with

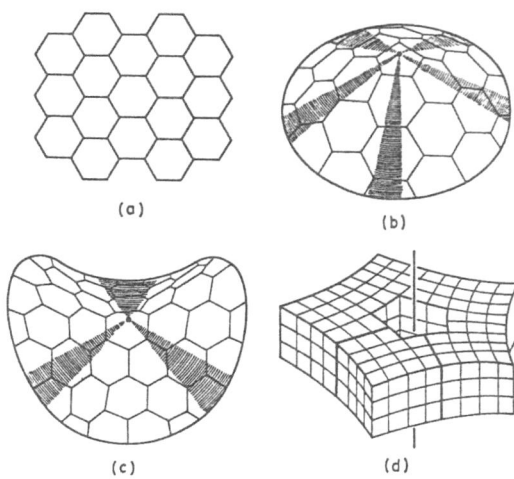

(a)

(b)

(c)   (d)

Fig. 5.5. Disclinations can be introduced into the plane hexagonal net (**a**) by removing or adding a 60° wedge, as in (**b**) and (**c**) respectively. A similar procedure in a 3D lattice is illustrated in (**d**)

respect to that present in the ordered reference system, i.e., the initial regular tessellation in spherical space.

5.6 Mapping from S^3 to E^3

We now turn to the mapping of the patterns of interest from S^3 to real space, E^3. Ideally, the mapping we should like to have is one which preserves lengths, since it is an experimental fact that amorphous solids do not exhibit much spread in the first-neighbour distances (i.e., bond lengths). A length-preserving mapping is called *isometric*. Unfortunately, such a mapping from S^3 to E^3 is not possible, because the two spaces have different curvatures. *Kléman* and *Sadoc* [5.3, 15] have therefore suggested other mapping procedures, namely, (i) star mapping, and (ii) mapping by a disclination procedure.

The 2D analogue of star mapping is shown in Fig. 5.6. Essentially it is like peeling an orange and flattening the peels to cover a planar region. In the process, severe distortions are introduced into the tiling. While the original tessellation of S^2 is regular, the flattened one is far from regular. There are bond length distortions as well as alterations in the coordination number, especially at vertices near the cuts. The figure also illustrates another obvious but important point: the peels of a single orange (of finite radius!) can cover only a portion of E^2.

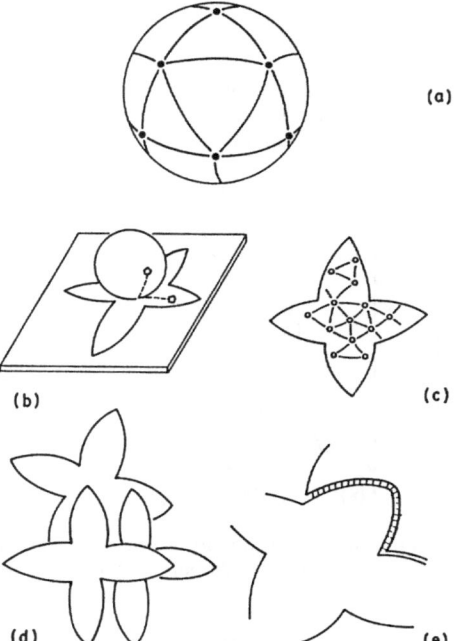

(a)

(b)   (c)

(d)   (e)

Fig. 5.6. (a) is a regular tessellation of S^2. To cover E^2 with the pattern on S^2, the latter must be peeled as in (b). This then leads, among other effects, to modifications in the coordination numbers at different vertices, as in (c). An attempt to cover the plane with peels is shown in (d). Overlaps may be avoided as in (e)

To cover the entire plane, keeping the basic average bond length fixed, an infinite number of oranges must be peeled. Even then there is a problem, as the peels do not mesh smoothly with each other. However, voids and overlaps could be avoided by suitable local distortions, facilitating sewing up to make a "carpet of orange peels". The pattern on the carpet will naturally be a highly distorted version of the original regular tessellation. A high density of topological defects will be present. Moreover, the pattern on such a carpet is far from unique.

Analogous steps can be visualized concerning the mapping of tessellations of S^3 onto E^3, although the steps are more involved [5.15]. Instead of 2D peels, one now has 3D lobes, as in Fig. 5.7. Any such star mapping will necessarily have only a finite number of vertices (or atoms, if atoms are assumed to be placed on the vertices), because each of the tessellations of S^3 has a finite number of vertices. To produce a space-filling structure in E^3 one must use an infinite number of such star-mappings which must then be meshed together as best as one can. Clearly, there is no unique way of doing this, but one could certainly aim at maximizing the density and minimizing the strain.

Figure 5.8 shows some results for the pair correlation function $g(r)$ obtained by *Sadoc* [5.16] in this fashion[1]. For the initial tessellation (or regular polytope) one has, obviously, a δ-function distribution as in the case of single crystals

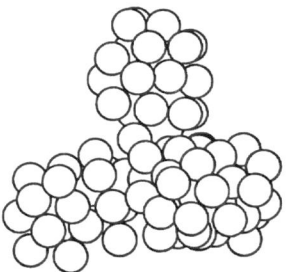

Fig. 5.7. View of the tessellation $\{3, 3, 5\}$ of S^3 after star mapping to E^3. (The spheres represent spherical atoms placed at the vertices of the tessellation)

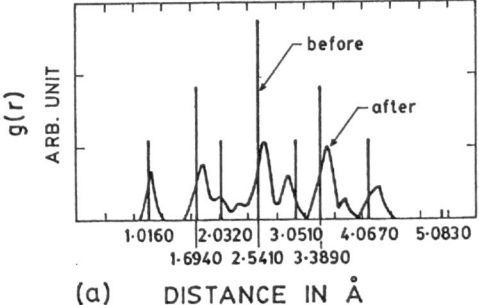

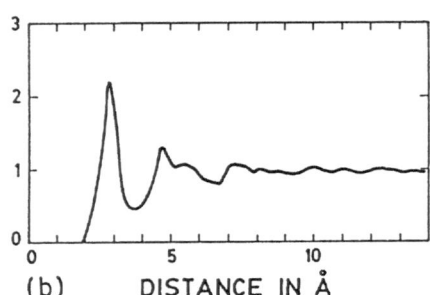

Fig. 5.8a, b. $g(r)$ for $\{3, 3, 5\}$ and its projection. (**a**) shows the function for a single polytope (120 atoms) before and after star mapping. The distribution for a 651-atom model synthesized from several star maps is shown in (**b**)

---

[1] The pair correlation function $g(r)$ is related to the function $G(r)$ in (2.3) by $G(r) = \delta(r) + g(r)$.

[5.17]. The irregular curve in Fig. 5.8a shows the distribution obtained upon peeling the polytope along four "longitudes". There is a shift in the positions of the peaks as well as a broadening, as expected. The curve in Fig. 5.8b shows $g(r)$ for a 651-atom cluster obtained by meshing together several projected structures. The result compares favourably with that for the DRP built by *Bernal* [5.18], but the density, however, is somewhat lower.

5.7 Defects and Star Mapping

We have already noted that when a spherical tessellation is 'peeled' and flattened, there occur (i) a slight spread in the bond distances and (ii) changes in the local topology in the neighbourhood of the cut surfaces (Fig. 5.6). Both these contribute to the "amorphization" of the structure. In particular, the cut surfaces lead to the formation of defects, which arise when one tries to sew together the various 'peels'. Consequent to such a sewing process, the coordination numbers of some of the vertices become anomalous, i.e., acquire values such as 14, 15, . . . etc. Frank–Kasper chains are generated thereby.

5.8 Mapping by Disclination Procedure

We have seen that disclinations may be introduced by cutting bonds and adding (or removing) a wedge of material between the two lips of the cut. Adding a wedge of material in spherical space *decreases* the curvature. Thus, introducing a positive disclination into a spherical tessellation achieves three things: (i) It introduces a defect, (ii) it decreases the curvature, and (iii) it increases the number of vertices (i.e., the number of atoms comprising the structure). For instance, *Sadoc* [5.19] has noted that introducing a particular pair of disclinations into $\{3, 3, 5\}$ increases the number of vertices from 120 to 168.

Given the above facts, one can visualize an alternative scheme to star mapping for generating amorphous structures from spherical tessellations (or regular polytopes). One adds the disclinations in iterative steps, a process referred to as a *disclination procedure*. Depending on how the iteration is done, various outcomes are possible, as indicated below.

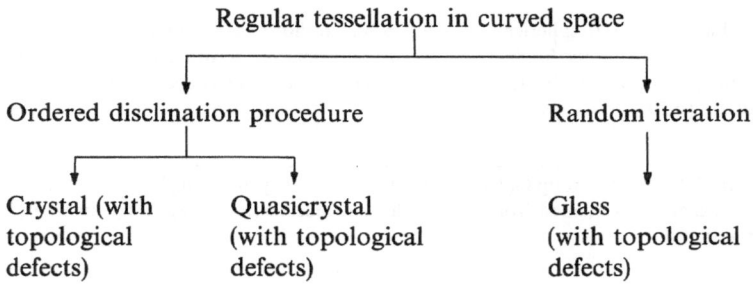

Such a scheme is interesting because it establishes a relationship between various structures. Incidentally, it is worth observing that there is no "cut and sew" prescription involved in the disclination procedure, in contrast to that of star mapping.

A brief description of the procedure [5.20] is as follows. The details are complicated, but the spirit of the procedure can be understood with the help of the following 2D example.

Consider S^2, and on it a geometric pattern which has the symmetry Y* of the full icosahedral group[2]. Such a pattern can be generated by starting with a suitably decorated fundamental triangle on S^2, and applying the operations of Y* to the triangle (Fig. 5.9).

Consider next the icosidodecahedron shown in Fig. 5.10a. This is not a Platonic solid (i.e., a regular polyhedron), but it is semiregular. It is made up of equilateral triangles and regular pentagons, and two triangles and two pentagons are linked to each vertex. A portion of the icosidodecahedron is now sliced and used to decorate the fundamental triangle on S^2 (the details are involved and may be found in [5.20]). Applying the operations of Y* to this decorated triangle, a new polyhedron P_1 of Fig. 5.10b is obtained. There are clearly many (topological) changes compared to the initial polyhedron P_0 of Fig. 5.10a, and we label these as defects. They are of two types. First, some triangles have been changed to pentagons, and second, some pentagons have been transformed to hexagons. The iteration $P_0 \rightarrow P_1$ has thus resulted in several disclinated polygons.

The iteration $P_1 \rightarrow P_2$ proceeds along similar lines. A piece of P_1 is used to decorate the fundamental triangle, and P_2 is then obtained via the action of $\dot{Y}*$. In this way, one can go through successive iterations. The procedure bears some

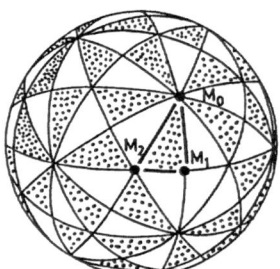

Fig. 5.9. Pattern of 120 spherical triangles generated from the fundamental triangle $M_0 M_1 M_2$ by the action of Y* [5.20]

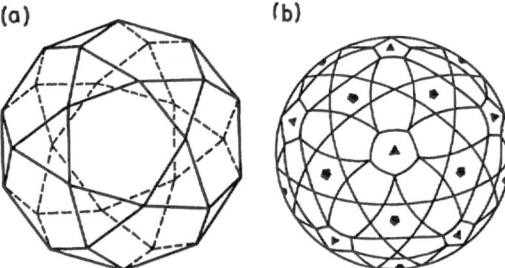

Fig. 5.10. (a) shows the icosidodecahedron. (b) shows the polyhedron obtained by disclinating (a). For convenience, the latter polyhedron has been projected onto S^2

---

[2] Just as the full rotation group O(2) is obtained from SO(2) by including inversion, the full group Y* with 120 elements is obtained from the 60 element icosahedral group Y by including improper transformations.

broad similarities to the inflation procedure for generating Penrose tilings (Sect. 2.2).

Sadoc and *Mosseri* [5.20] have used the 3D extension of the above procedure to study the consequences of disclinating $\{3, 3, 5\}$. They first note that for every iteration $P_i \to P_{i+1}$ (where P_i and P_{i+1} denote the polytopes in two successive steps), there are two options available, each with its own characteristic defect set. Let the two transformation possibilities be labelled as a and b. The n iterations involved in going from P_0 ($= \{3, 3, 5\}$) to P_n can then be characterized by an n-letter string such as *abbabaa* . . . . The complexity of the structure is encoded in the information content of the n-letter word. The different iteration schemes indicated in the tree given earlier correspond to different words.

Let us illustrate the meaning of such iterations by considering a particular example. Recall that all the vertices in $\{3, 3, 5\}$ have only icosahedral coordination, i.e., $Z = 12$. But when $\{3, 3, 5\}$ is disclinated, vertices with $Z = 14$ and 16 also appear. Let $\mathcal{N}^{(i)} = (N_{12}^{(i)}, N_{14}^{(i)}, N_{16}^{(i)})^T$ denote the vector whose components are the total number of $Z = 12$, $Z = 14$ and $Z = 16$ sites in the polytope P_i. It turns out that the transformation can be represented by the matrix

$$\underset{\sim}{T} = \begin{pmatrix} 13 & 12 & 12 \\ 0 & 3 & 4 \\ 5 & 6 & 8 \end{pmatrix}. \tag{5.4}$$

In other words, the iteration $P_i \to P_{i+1}$ yields

$$\mathcal{N}^{(i+1)} = \underset{\sim}{T} \mathcal{N}^{(i)}. \tag{5.5}$$

The results of disclinating $\{3, 3, 5\}$ repeatedly using (5.4) are summarized in Table 5.2. The appearance of vertices with $Z = 14$ and $Z = 16$ indicates the presence of disclinations or Frank–Kasper chains in P_i ($i > 1$). The asymptotic value of the average $\bar{Z}$ so obtained ($= 40/3$) may be compared with the "ideal value" of 13.397 . . . [5.21] for Frank–Kasper structures.

Table 5.2. Characterization of the polytopes P obtained from $\{3, 3, 5\}$ by repeated application of the disclination transformation (5.4). N_z is the number of sites with coordination number Z. N is the total number of vertices, T the total number of tetrahedral cells and $\bar{Z}$ is the average coordination number

Stage of iteration	N_{12}	N_{14}	N_{16}	N	T	$\bar{Z}$
P_0	120	0	0	120	600	12
P_1	1560	0	600	2160	12000	13.111
P_2	27480	2400	12600	42480	240000	13.2999
P_3	537240	57600	252600	847400	4800000	13.328
P_4	10706520	1183200	5052600	16942320	96000000	13.3325
P_5	214014360	23760000	101052600	338826960	1920000000	13.3332

5.9 Decoration

The polytopes $\{3, 3, 5\}$ and $\{5, 3, 3\}$ (or rather, the corresponding regular tessellations of S^3) generate respectively DRP and CRN structures. Since the number of *regular* polytopes is rather small (Table 5.1), one might wonder whether this places any restriction on the types of amorphous structures one can generate from the tessellations of curved space. It turns out that *decoration* makes numerous variations possible. In the case of periodic lattices, decoration is the process of attaching an identical pattern to each lattice site (or bond). In this manner, one can progress from unit cells with one atom to cells with many atoms. For instance, we can obtain the face-centred cubic and body-centred cubic lattices by suitably decorating the simple cubic lattice.

Decorated polytopes are obtained in a similar fashion. One example is the "polytope 240" obtained from $\{3, 3, 5\}$ by introducing a new vertex at the centre of one of the 5 tetrahedra surrounding each edge [5.15]. There are thus $(120 + 600/5) = 240$ vertices in the decorated polytope. Placing Si atoms at these 240 vertices and projecting the structure onto E^3, *Sadoc* and *Mosseri* [5.15] obtained a model for amorphous silicon (a-Si). This model has only even-membered rings, in contrast to a ball-and-stick model constructed earlier by *Polk* [5.22]. (The latter model has been referred to in Chap. 2, and resembles Fig. 2.9b. A dominant feature of the model is the frequent occurrence of 5-membered rings.) The model of Sadoc and Mosserri, on the other hand, is akin to that of *Connell* and *Temkin* [5.23] which avoids odd-membered rings. (It is believed that, for reasons connected with overall charge neutrality, the Connell–Temkin model is an appropriate one for a-GaAs.)

Decoration of curved space tessellations and subsequent projection onto E^3 reproduces a number of other amorphous structures in addition to a-Si. For instance, *Sadoc* and *Mosseri* [5.24] have proposed a model for the common metallic glass $Fe_{80}B_{20}$ by decorating $\{5, 3, 3\}$ along its edges with trigonal prismatic units (Fig. 5.11). There are several other examples as well [5.25].

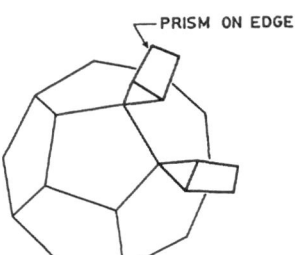

PRISM ON EDGE

Fig. 5.11. The polytope $\{5, 3, 3\}$ is a packing of dodecahedra, three of them joined at each edge. Trigonal prismatic units are placed on the edges as shown. By placing metal and nonmetal atoms suitably in the prisms, one can reproduce the structure of a-$Fe_{80}B_{20}$ after projection on to E^3

5.10 Defects in the CRN

We have already noted that defects in the DRP are Frank–Kasper lines, i.e., lines on which vertices have anomalous coordination ($Z \neq 12$). Alternatively, these lines may be viewed as chains of bonds, each bond of which is shared by an anomalous number (4, 6, 7, . . .) of tetrahedra. It is interesting to ask what the nature of the line defect is in the CRN. We note first that the CRN is derived from the tessellation $\{5, 3, 3\}$ [5.15]. To study the nature of the defect in the CRN we must therefore disclinate $\{5, 3, 3\}$. When this is done, it turns out that the disclinated structure has nonpentagonal rings as well, in contrast to $\{5, 3, 3\}$, which has only pentagonal rings. Line defects analogous to the Frank–Kasper lines are now identified by threading lines through nonpentagonal rings; each line passes through rings with a given number of sides (4 or 6 or 7 . . .). The generation of such nonpentagonal rings involves once again the removal or addition of wedges, so that these lines also qualify as disclinations.

It is known [5.26, 27] that the polytopes $\{3, 3, 5\}$ and $\{5, 3, 3\}$ share the same symmetry group. Recall also that the two are duals. Therefore, when the topological defect classification method is applied, one may expect to find correspondences in the defect structures of the two polytopes, and hence in the defect structures of the projected Euclidean structures (the DRP and the CRN) derived respectively from these polytopes. There is indeed such a correspondence. Instead of a chain of bonds each shared by 4 or 6 or 7 . . . tetrahedra, we have a line threading through rings having 4 or 6 or 7 . . . edges.

In general, the defects in an amorphous structure can be discerned by carefully identifying the original polytope (which may be a regular polytope or a decorated one), and then disclinating it. (Recall also Fig. 5.10 in which a non-platonic polyhedron was disclinated.)

5.11 Amorphous Structures by Projection of Hyperbolic Tilings

In their original paper, *Sadoc* and *Kléman* [5.3] pointed out that tessellations of both S^3 as well as H^3 were possible candidates for generating amorphous structures in E^3. However, subsequent work has tended to concentrate mainly on the mapping of tessellations of S^3, rather than those of H^3. One reason is that the latter often lead to structures with too large a value for the average coordination number, $\bar{Z}$. (See Appendix C.) Nevertheless, *Kléman* [5.28] has pointed out that there are advantages in considering tilings on H^3 as well. In particular, *Kléman* and *Donnadieu* [5.29] note that mapping from H^3 is a way of generating unusual defects like *disvections*.

5.12 Polymers and Polytopes

In Chap. 2 we classified amorphous materials into three broad categories: DRP structures, CRNs and polymeric structures. We have described how structures belonging to the first two categories can be generated by projecting suitable tessellations of S^3 onto E^3. One may ask whether a similar technique is applicable to polymeric structures. *Kléman* [5.30] has recently suggested that it is, provided one starts with an ordered pattern on S^3 made up of chiral strands or strings. Such generalizations of the original polyhedral tilings of S^3 appear to suggest new ways of generating a whole range of complex structures in E^3. An example is the disclinated "double twist" structure that has been suggested for the blue phase of cholesterics [5.31].

5.13 Quasicrystals by the Projection Method

We turn now to the application of the projection method to quasicrystals. Historically, incommensurate structures were discovered first, and it was soon recognized that they could be viewed as projections of periodic structures in higher-dimensional Euclidean spaces [5.5, 6]. Subsequently, after the discovery of Penrose tilings, 2D and 3D tilings of this kind have been shown to be projections of higher-dimensional periodic structures [5.32, 33]. The discovery of quasicrystallinity [5.34] has led to renewed interest in the projection method, and several new aspects have been clarified recently [5.7–10, 35–37].

5.13.1 Generation of the Penrose Chain

We recall that the Penrose chain (Sect. 2.3) is characterized by two lengths L and S in the ratio $\tau:1$, and that these "tiles" occur in a Fibonacci sequence. The steps involved in deriving this chain by the projection technique are as follows [5.8]:

i) Construct a square lattice with lattice spacing equal to unity. Select a lattice point and label it as the origin 0. (Fig. 5.12).

ii) Through 0, draw a line making an angle θ with the x-axis such that $\tan\theta = \tau^{-1}$. We will refer to the space represented by this line as ξ. The space ξ contains no lattice point other than 0.

iii) Select a unit cell having 0 as one of its corners, e.g., as in Fig. 5.12. Let the open region $\{(x, y)|0 < x < 1, 0 < y < 1\}$ be designated γ_2.

iv) Generate a strip $\mathscr{S}$ by moving γ_2 along ξ without rotation.

v) Consider the lattice points contained entirely in the strip $\mathscr{S}$. (There is a unique and continuous zig-zag line made up of the bonds contained entirely in $\mathscr{S}$.) Project the lattice points in $\mathscr{S}$ onto ξ (Fig. 5.12). These projections form the Penrose chain.

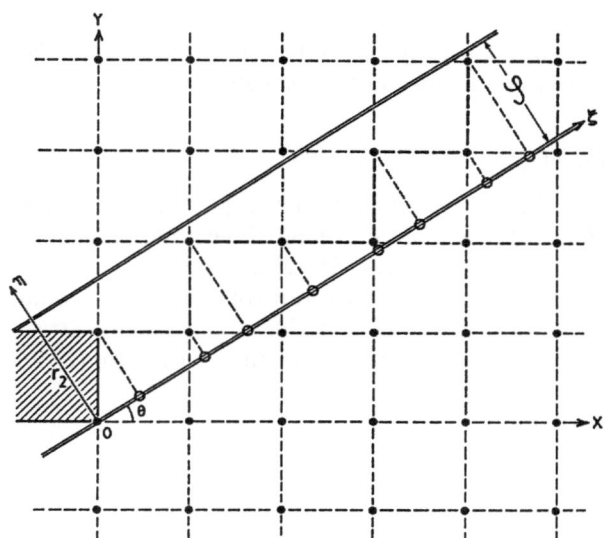

Fig. 5.12. Realization of a quasiperiodic chain by projection from a square lattice. The 1D sequence of points is the set of projected images of the points lying within the strip $\mathscr{S}$

5.13.2 Varying the Choice of the Unit Cell

It is well known that the unit cell of a crystal can be chosen in different ways. Similarly, in the foregoing construction any larger square of the lattice $\mathbb{Z}^2$ can be chosen to take the place of γ_2. A larger unit cell will lead to a wider strip $\mathscr{S}$, and a denser set of projected points on ξ. However, the new chain will be merely a deflation of the (infinite) chain obtained using γ_2.

5.13.3 Role of the Slope

In Fig. 5.12, the particular choice made for the slope (i.e., $\tan \theta = \tau^{-1}$) obviously has an important influence on the final outcome. In general, the slope can be shown to be the ratio of the relative abundances of the two types of tiles occurring in the projected structure (the "stoichiometry"). As θ is increased from zero, one generates a variety of structures featuring two tiles. Whenever the slope is a rational number, a long-period structure results, while when it is irrational a nonperiodic structure is projected out. Indeed, we can generate all the structures illustrated in Fig. 2.3: these are really based on various rational approximations to τ.

5.13.4 Effect of Translating $\mathscr{S}$ Laterally

The strip $\mathscr{S}$ may be relocated by moving it in the transverse direction. It now encloses a new set of lattice points, whose projections onto ξ yield another 1D quasiperiodic tiling. This is the same local isomorphism class (Sect. 2.3.3) as the original chain. If $\mathscr{S}$ is translated by a *lattice* vector of the square lattice, then the resulting projection is just a translated version of the original one.

5.13.5 Role of the Orientation of ξ

In the scheme discussed in Sect. 5.13.1, the line ξ and the strip $\mathscr{S}$ have the same slope. This is necessary in order to obtain a Penrose chain. If the orientation of $\mathscr{S}$ is retained but that of ξ alone is altered, a new quasiperiodic chain would result. The tile sequence would be the same as before, but the ratio of the tile *sizes* would no longer be equal to τ. Thus, the orientation of $\mathscr{S}$ decides the "stoichiometry", while the orientation of ξ defines the relative sizes of the tiles; recall also (2.19).

5.14 Generalization

We consider now the generalization of the above method. The idea is to generate a quasiperiodic tiling of a p-dimensional space ξ by projection from the (hypercubic) lattice $\mathbb{Z}^n$ in the space E^n. The steps are as follows.

   i) Let $(\varepsilon_1, \ldots, \varepsilon_n)$ denote the orthonormal basis vectors of the hypercubic lattice $\mathbb{Z}^n$. Consider the unit cube γ_n (the analogue of γ_2).
  ii) Select a p-dimensional subspace ξ (a hyperplane) that does not contain any point of the lattice except the origin.
 iii) Shift γ_n along ξ to generate the "strip" "$\gamma_n + \xi$", denoted by $\mathscr{S}$.
  iv) Select all the lattice points of $\mathbb{Z}^n$ which fall inside $\mathscr{S}$. This procedure selects a p-dimensional zigzag surface akin to the zigzag line of Fig. 5.12. This surface is a union of p-facets (the p-dimensional analogue of an edge).
   v) Project the p-dimensional surface generated in (iv) onto ξ. This gives the required p-dimensional tiling.

The generation of (2D) Penrose tilings is a special case of the above procedure [5.37]. Since the essential feature of the Penrose tiling is its five-fold symmetry, we first search for the simplest hypercubic lattice with five-fold symmetry. This is the lattice $\mathbb{Z}^5$ in E^5. Let Δ denote a principal diagonal of the lattice $\mathbb{Z}^5$. (Δ is like the [111] axis of the simple cubic lattice in E^3, i.e., of $\mathbb{Z}^3$.)

Let G be the point group of $\mathbb{Z}^5$. Its character table is given in Table 5.3. Under the action of this group, the space E^5 breaks up into three invariant subspaces: two planes P_1 and P_2 and a line L. These are the spaces of the irreducible representations Γ_2, Γ_2' and Γ_1 respectively. The strip is defined by $\mathscr{S} = \gamma_5 + P_1$, where γ_5 is the open unit cube of E^5, and P_1 plays the role of ξ. When all the points of $\mathbb{Z}^5$ in $\mathscr{S}$ are projected onto P_1, a Penrose tiling results. Variations are obtained by translating $\mathscr{S}$ by a vector t, where t is any vector of the space $(P_2 + \Delta)$.

Table 5.3. Character table for the point group of the cubic lattice in five dimensions. Here E is the identity operation, C_5 is a rotation by $(2\pi/5)$ and m, a mirror operation. As usual, $\tau = (\sqrt{5}+1)/2$

Irreducible representation	E	$2C_5$	$2C_5^2$	$5m$
Γ_1	1	1	1	1
Γ_1'	1	1	1	-1
Γ_2	2	τ	$-(1+\tau)$	0
Γ_2'	2	$-(1+\tau)$	τ	0

The generation of 3D icosahedral tilings also follows a similar procedure. The relevant higher-dimensional lattice in this case is $\mathbb{Z}^6$ in E^6. This lattice is invariant under the icosahedral point group whose character table is given in Table 5.4. Under the action of this group, E^6 decomposes into two invariant subspaces ξ and ξ', associated respectively with the irreducible representations Γ_3 and Γ_3'. The lattice $\mathbb{Z}^6$ has 20 different 3-facets generated by combinations of

Table 5.4. Character table for the icosahedral permutation group. C_n is a rotation by $(2\pi/n)$

Irreducible representation	E	$12C_5$	$12C_5^2$	$20C_3$	$15C_2$
Γ_1	1	1	1	1	1
Γ_3	3	$1+\tau$	$-\tau$	0	-1
Γ_3'	3	$-\tau$	$1+\tau$	0	-1
Γ_4	4	-1	-1	1	0
Γ_5	5	0	0	-1	0
Γ_6	6	1	1	0	-1

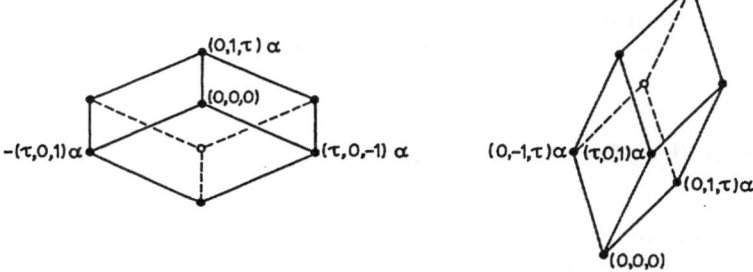

Fig. 5.13. The two rhombohedral unit cells used to build the icosahedral quasicrystalline packing. The ratio of the volumes of the oblate rhombohedron (*left*) to the prolate one (*right*) is τ

basis vectors like $(\varepsilon_1, \varepsilon_2, \varepsilon_3)$, etc. These facets are projected onto ξ in the form of the two rhombohedra shown in Fig. 5.13. In other words, just as the facets of the square lattice are projected as the L and S tiles, the facets of $\mathbb{Z}^6$ are projected as the "thick" and the "thin" rhombohedra. The strip $\mathscr{S}$ is generated by shifting the unit cube γ_6 along ξ. All the 3-facets entirely contained in $\mathscr{S}$ are projected onto the three-dimensional space ξ to obtain the desired icosahedral tiling.

5.15 Some Comments on the Projection Method

Projections of (a slice of) only the hypercubic lattice $\mathbb{Z}^n$ in E^n have been considered in the foregoing. One could also start with a noncubic lattice, or even a decorated lattice. These would lead to a rich variety of structures.

A feature of the projection method is that structures generated by it are self-similar. *Socolar* and *Steinhardt* [5.35] have pointed out that the Generalized Dual Method (GDM; Sect. 2.2) is capable of generating more general tilings. The most general 3D icosahedral packing is obtained by means of the GDM when one takes six grids of arbitrary spacing, each grid oriented normal to one of the six independent axes of five-fold symmetry of an icosahedron. A subset of such structures is obtained when one uses Fibonacci hexagrids, i.e., when the spacing pattern in each grid follows a Fibonacci sequence. Structures obtained by projection from the 6D hypercubic lattice in turn constitute a subset of the duals of Fibonacci hexagrids.

Katz and *Duneau* [5.8] mention that more general tilings in 1D may be derived with a strip obtained by moving the unit cell along a curved or undulating line, while maintaining the projection on the straight line ξ. The higher dimensional analogues of this procedure have yet to be explored systematically.

Finally, an important aspect of the Sadoc–Kléman method of projection from curved space is that it leads to topological defects in random structures. The projection technique for quasicrystals currently under discussion does not emphasize topological defects. The question of possible topological defects in quasicrystals is an intricate one. Some results (based on the projection properties described above) have been obtained recently in this regard [5.38, 39].

5.16 Miller Indices for Quasicrystals

Concepts analogous to those of crystal planes (hkl) and crystal directions $[hkl]$ exist in the case of quasicrystals. To be specific, let us consider a 3D icosahedral structure. Let $\{\varepsilon_i\}$ $(i = 1, \dots, 6)$ denote the basis vectors of $\mathbb{Z}^6$, and let $\{e_i\}$ be their projections onto the 3D space ξ. Consider the vector $\boldsymbol{d} = \sum_i n_i e_i$, where $\{n_i\}$

are integers. The direction defined by d describes a quasilattice direction, and is indexed by the set $[n_1, \ldots, n_6]$. The latter is the analogue of $[hkl]$. The quasilattice plane $(n_1, \ldots, n_6)$ is the plane perpendicular to d, and is the generalization of (hkl). The indices $(n_1, \ldots, n_6)$ are thus generalized Miller indices. As in the case of ordinary crystals, one may associate the spots in the diffraction pattern with vectors belonging to the lattice $\mathbb{Z}^{6*}$ which is reciprocal to $\mathbb{Z}^6$.

Sometimes there can be redundancies in the indices. In the case of hexagonal close-packed lattices, for example, one uses for convenience the set of *four* indices $(hkjl)$, with $j = h + k$. Likewise, there are redundancies in the indices in the case of some quasicrystals. The most common example is that of the Penrose tiling itself. Although one uses five Miller indices, four are sufficient on account of a redundancy. However, the use of five indices clearly exhibits the relationship of the structure to $\mathbb{Z}^5$.

5.17 Diffraction Patterns of Quasicrystals

One might expect that it would be difficult to compute the diffraction pattern of a quasicrystal on account of its lack of periodicity. While this is true in general, the problem is simpler for quasicrystals obtained by projection.

We illustrate the results by considering the 1D case [5.10]. Going back to Fig. 5.12, let ξ and η denote coordinates respectively parallel to the strip and normal to it. For convenience, one chooses the strip $\mathscr{S}(\eta)$ to be symmetric about the ξ-axis, i.e.,

$$\mathscr{S}(\eta) = \begin{cases} 1 & \text{if} \quad |\eta| < w/2 \\ 0 & \text{otherwise ,} \end{cases} \tag{5.6}$$

where w is the width of the strip. The chain projected by the strip is given by the sequence [5.40]

$$x_N = \left(N + \frac{1}{\tau} \left[\frac{N}{\tau} + \frac{1}{2} \right] \right) \sin \theta . \tag{5.7}$$

(5.7) corresponds to the special case $\alpha = 0$, $\beta = (1/2)$, $a = \sin \theta$ and $\varrho = \sigma = \tau$ of the general quasiperiodic sequence we discussed earlier in (2.19).

Let us write the density distribution $L_0(x, y)$ in the square lattice as

$$L_0(x, y) = \frac{1}{4\pi^2} \sum_{j,l} \delta(x - j)\,\delta(y - l) , \tag{5.8}$$

where the sum is over all integers j, l. Then the distribution $L(\xi, \eta)$ with respect to the rotated axes ξ, η (Fig. 5.12) is given by

$$L(\xi, \eta) = L_0(\xi \cos \theta - \eta \sin \theta, \xi \sin \theta + \eta \cos \theta) \ . \tag{5.9}$$

The distribution $\varrho(\xi)$ projected along ξ is then formally

$$\varrho(\xi) = \int d\eta \ \mathscr{S}(\eta) L(\xi, \eta) \tag{5.10}$$

with $\mathscr{S}(\eta)$ as in (5.6). Using the formula for the Fourier transform of products, we obtain from (5.10) the Fourier transform

$$\tilde{\varrho}(q) = \frac{1}{2\pi} \int dp \ \tilde{\mathscr{S}}(-p) \tilde{L}(q, p) \ . \tag{5.11}$$

Here the transform of the strip function (5.6) is given by

$$\tilde{\mathscr{S}}(p) = w \left(\sin \frac{pw}{2} \right) \Big/ \left(\frac{pw}{2} \right) \ . \tag{5.12}$$

Further, using the identity

$$\sum_l \exp(2\pi i l \alpha) = \sum_l \exp\left[2\pi i l(\alpha - m)\right] = \delta(\alpha - m) \ ,$$

where m is any integer, the transform of $L(\xi, \eta)$ becomes

$$\tilde{L}(q, p) = \delta(q - 2\pi(n \cos \theta + m \sin \theta))\delta(p - 2\pi(-n \sin \theta + m \cos \theta)) \ , \tag{5.13}$$

where m and n are arbitrary integers. Therefore

$$\tilde{\varrho}(q) = \frac{1}{2\pi} \tilde{\mathscr{S}}(2\pi(n \sin \theta - m \cos \theta)) \delta(q - 2\pi(n \cos \theta + m \sin \theta)) \ . \tag{5.14}$$

Recalling that

$$\sin \theta = 1/\sqrt{1 + \tau^2} \ , \qquad \cos \theta = \tau/\sqrt{1 + \tau^2} \ , \tag{5.15}$$

we see that diffraction spots occur whenever

$$q = \frac{2\pi}{\sqrt{1 + \tau^2}}(m + n\tau) \ . \tag{5.16}$$

This is to be compared with the crystalline case where (for unit lattice spacing)

$$q = 2\pi m \ .$$

As the Penrose chain has two length scales, there are two Miller indices; as a result, the diffraction spots densely fill the line representing q-space. However, owing to the modulation factor $\tilde{\mathscr{S}}(-p)$, all spots are not equally bright; spots with high Miller indices are weak. It is noteworthy that, unlike the crystalline case, high index reflections do not necessarily occur far out in reciprocal space. They could in fact be quite close to the origin.

A brief comment now about the celebrated five-fold symmetry of the diffraction spots of the quasicrystalline alloy Al-Mn (Fig. 5.14): In principle one should expect spots over the whole of q-space. However, owing to the filtering action of the strip function, not all spots are equally bright, and the observed pattern exhibits the underlying orientational symmetry. Indeed, this is fortunate, for this intriguing structure might otherwise never have revealed itself.

Fig. 5.14. Electron diffraction pattern of the icosahedral phase of Al-14 Mn (as quenched) alloy, courtesy Dr. V. S. Raghunathan

5.18 Incommensurate Crystals

As already remarked, quasicrystals are not the only systems to exhibit quasiperiodicity. Even prior to their discovery, other structures with quasiperiodicity were known. However, all of them had orientational symmetries consistent with the allowed crystallographic point groups. Such systems are comprehensively referred to as incommensurate crystals. We shall see presently how such systems too can be viewed as projections of higher-dimensional periodic structures. Physical examples of incommensurate structures include helical spin systems like Tm, layered compounds like TaS_2 with charge-density waves, and ionic crystals like Na_2CO_3 with displacive modulations. For simplicity, we shall consider the case of displacive modulations.

Consider an ordinary crystal with a translational lattice Λ. Let $(\boldsymbol{a}_1, \boldsymbol{a}_2, \boldsymbol{a}_3)$ denote the basis vectors of Λ and $(\boldsymbol{a}_1^*, \boldsymbol{a}_2^*, \boldsymbol{a}_3^*)$ those of the reciprocal lattice Λ^*.

We suppose that there are s atoms in the unit cell with equilibrium positions

$$x(l, k) = x(l) + x(k) , \quad (k = 1, \ldots , s) \tag{5.17}$$

where l labels the unit cell and k is the index of the atom in the unit cell. If the crystal is transformed into a modulated structure, the positions are given by

$$X(l, k) = x(l) + x(k) + f(k) \exp [i q \cdot x(k)] , \tag{5.18}$$

where q is the wavevector of the modulation wave. For simplicity we assume that there is only one such wave.

Now any vector $G \in \Lambda^*$ can be expressed as

$$G = h_1 a_1^* + h_2 a_2^* + h_3 a_3^* , \tag{5.19}$$

where (h_1, h_2, h_3) are integers. In terms of the same basis, the vector q in (5.18) can be written as

$$q = f_1 a_1^* + f_2 a_2^* + f_3 a_3^* , \tag{5.20}$$

where (f_1, f_2, f_3) are real numbers. An alternative representation is

$$q = G + h_4 a_4^* . \tag{5.21}$$

Here G is given by (5.19), h_4 is an integer and a_4^* is defined by

$$a_4^* = \sigma_1 a_1^* + \sigma_2 a_2^* + \sigma_3 a_3^* . \tag{5.22}$$

Incommensurability means that at least one of the σ_i's in (5.22) is irrational. If this is the case, then the crystal described by (5.18) lacks translational periodicity (recall Fig. 2.1c). *Janner* and *Janssen* [5.5] and *de Wolff* [5.6] showed that a structure described by (5.18), though nonperiodic in the usual sense, can be regarded nevertheless as a periodic crystal in 4D space.

The actual realization of the 4D supercrystal is a bit involved. Using $(a_1^*, a_2^*, a_3^*, a_4^*)$, one first generates four new vectors defined by

$$b_1 = a_1^* , \quad b_2 = a_2^* , \quad b_3 = a_3^* , \quad b_4 = a_4^* + e , \tag{5.23}$$

where e is a unit vector perpendicular to E^3 (the space of Λ^*). Using $(b_1, \ldots, b_4)$, a lattice Σ^* is generated in E^4. Let Σ denote the lattice reciprocal to Σ^*. It then turns out that Σ is a periodic representation in E^4 of the modulated crystal defined by (5.18, 20 and 21). The structure in E^3 defined by the above equations is obtained by intersecting the 4D periodic lattice Σ with the hyperplane E^3 normal to e, i.e., by projecting Σ onto E^3 along e.

It is noteworthy that, unlike a quasicrystal, an incommensurate crystal can be constructed by starting with a reference (3D, crystalline) lattice, recall (5.17, 19). For this reason, the diffraction pattern exhibits an underlying lattice structure;

the effects of incommensurate modulation manifest themselves as satellite spots around the basic reflections characterized by G (Fig. 5.15). The spots are indexed as usual, i.e., using the vectors of Σ^*.

Janner and Janssen have considered also the possibility of more than one modulation being present. If there are d such modulation waves then the superstructure is of dimensionality $(3 + d)$. *Ruelle* [5.41, 42] has observed that "a crystal with five spatial frequencies, i.e., two independent, incommensurate modulations, could exhibit spatial turbulence".

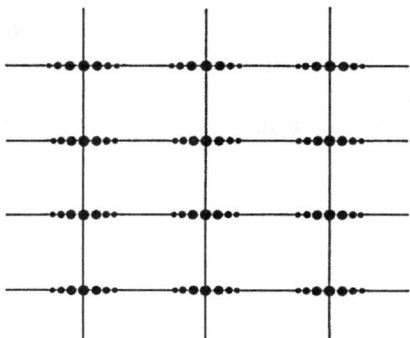

Fig. 5.15. Diffraction pattern of a 2D incommensurate structure. The "decorated lattice" form of the pattern may be noted

5.19 Summary

In this chapter we have seen how some Euclidean noncrystalline structures can be regarded as projections of ordered structures in other spaces. One point of view, promoted particularly by the French school, is that amorphous structures can be obtained from ordered structures in S^3 or H^3. In either case, the ordered structure from which one starts is chosen so as to have a local topology related to that of its Euclidean counterpart. If the mapping is achieved via a disclination procedure, then, depending on the iterations employed, one can arrive at a crystalline, a quasicrystalline, or an amorphous structure. An important feature of all structures obtained in this manner is the presence of topological line defects. For the DRP structure, this defect is the same as the familiar Frank–Kasper chain. Its nature in various other structures remains to be elucidated further.

The alternative approach, tailored specifically for quasicrystals and, to a certain extent, for incommensurate crystals, is to start with a higher-dimensional Euclidean structure, rather than a structure in a curved space of the same dimensionality. The higher-dimensional space group is chosen so as to have the same rotational symmetry as that of the structure one is interested in. Studies of structures projected onto 1D suggest that, in principle, this technique can be employed to derive long-period structures as well. There exists as yet no conclusive demonstration that amorphous structures can also be derived by this technique.

There are thus at least two distinct geometric routes for arriving at some of the noncrystalline structures. At present it is difficult to say which of them is superior. But it would appear that both deal with systems which have some sort of "internal degrees of freedom". These extra degrees of freedom are "subsumed" in some sense into the extra dimensions or the curvature of the space, respectively.

The fact that we can create various patterns by projection either from curved spaces or higher-dimensional Euclidean spaces does not mean that such structures necessarily occur in Nature. For that to happen, there must be an advantage in terms of energy minimization. One must therefore look beyond mere geometry to understand the existence of the various structures we have been discussing. This is taken up in the next chapter.

6. Beyond Simple Geometry

As already indicated, Nature does not rely on geometry alone to build structures. Energy considerations also play an important role, to which we now direct attention.

The simplest way of taking some account of energetics is via the Landau theory [6.1], introduced briefly in Chap. 3. Landau's objective was to provide a phenomenological description of the emergence of various types of structural order from a liquid. Two crucial (and related) concepts involved in the theory are symmetry breaking and the order parameter.

Over the years, the Landau theory has evolved considerably and has been applied in its various updated forms to diverse problems. This chapter is largely concerned with a study of the emergence of structural order, within the framework of the Landau theory.

6.1 Some Basics

As in Chap. 3, let us start with the free energy expansion for a system with a one-component order parameter, near $T = T_c$:

$$F = a(T - T_c)\psi^2 + c\psi^4 , \qquad a, c > 0 . \tag{6.1}$$

Earlier we have seen that below T_c, the system orders with a nonvanishing value ψ_0 for the order parameter, given by

$$\psi_0^2 = a(T_c - T)/(2c) \tag{6.2}$$

[this follows readily upon using (3.3)]. The variation of F with ψ at various temperatures and the T-dependence of ψ_0 are sketched in Fig. 6.1(i). The transition is continuous (or of second order, depending on one's preference for words).

Suppose a third-order term is present in F, so that

$$F = a(T - T_c)\psi^2 + b\psi^3 + c\psi^4 . \tag{6.3}$$

In this case there is a discontinuous change in the value of ψ_0 at the transition temperature, and the transition is of first order. Consequent to such a jump, the

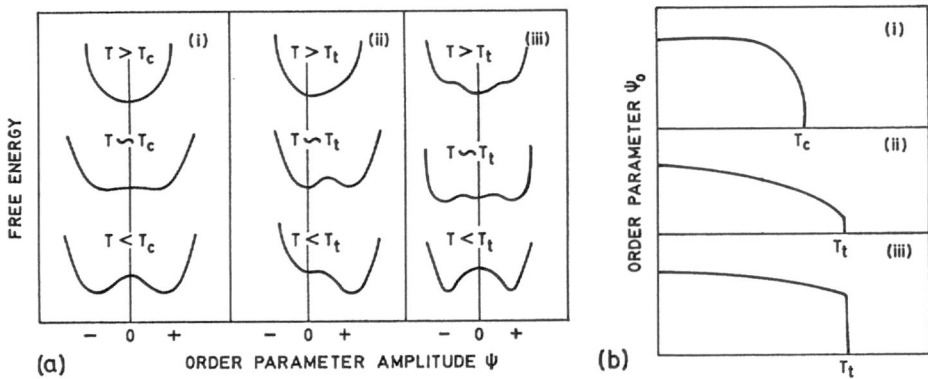

Fig. 6.1a, b. Schematic plots of $F(\psi)$ and $\psi_0(T)$ corresponding to Eqs. (6.1), (6.3) and (6.5). These three cases are identified by (i), (ii) and (iii) respectively

order parameter will in general not be small near the transition, and an expansion as in (6.3) is invalid. However, in many systems the jump is not large. One could then regard the transition as *nearly continuous* and apply the Landau theory as usual. In such an approach, the transition occurs at a temperature

$$T_t = T_c + \Delta , \quad \Delta = b^2/4ac , \tag{6.4}$$

above T_c. The forms of the free energy functions and the temperature dependence of the order parameter are sketched in Fig. 6.1(ii). The discontinuous nature of the transition is due to the third order term whose presence or absence is governed by symmetry. The transition associated with (6.3) is thus "symmetry driven".

First order transitions, not driven by symmetry, are also possible, for example when the free energy is given by

$$F = a(T - T_c)\psi^2 + c\psi^4 + e\psi^6 , \quad a > 0, \quad c < 0, \quad e > 0 ; \tag{6.5}$$

there is a discontinuous transition at $T_t \simeq T_c + c^2/4ae$. Some results for this case are sketched in Fig. 6.1(iii).

6.2 Landau Theory and Ordered Atomic Structures

6.2.1 Free Energy Expansion

When dealing with atomic arrangements, it is convenient to use the density function $\varrho(x, y, z)$, where $\varrho(x, y, z) \, dx \, dy \, dz$ is the probability of finding an atom in the volume element $dx \, dy \, dz$ in the neighbourhood of the point (x, y, z). If several atomic species are present, one must use separate density functions for each of the species. For simplicity, we shall here assume the system to be monatomic.

Consider a liquid characterized by a distribution $\varrho_0(x, y, z)$; in fact, ϱ_0 would be a constant. Let $\varrho = \varrho_0 + \delta\varrho$ characterize the distribution in the ordered state which emerges when the liquid is cooled (recall Table 3.2). If G and H denote the symmetry groups of the liquid and the broken symmetry state respectively, then ϱ_0 is left invariant by the operations of G $[= \mathrm{E}(3)]$ while ϱ is invariant under H. On the other hand, under the action of an element of G which is not contained in H, $\delta\varrho$ transforms as

$$\delta\varrho \rightarrow \delta\varrho'(x, y, z) = g\,\delta\varrho(x, y, z) = \delta\varrho[g^{-1}(x, y, z)] \; . \tag{6.6}$$

The distribution $\varrho' = \varrho_0 + \delta\varrho'$ is another possible realization of the state with symmetry H.

Following Landau [6.1, 2] we now write

$$\delta\varrho = \sum_{i=1}^{n} \eta_i^{\Gamma} \phi_i^{\Gamma} \; , \tag{6.7}$$

where ϕ_i^{Γ} are basis functions transforming according to the n-dimensional irreducible representation (IR) Γ of G, and $\{\eta_i^{\Gamma}\}$ are the expansion coefficients. The physical content of (6.7) is that the ordered phase is formed by the freezing of a *particular density fluctuation* of the liquid, the fluctuation being characterized by the IR Γ of G. This labelling is analogous to the way one characterizes lattice vibrations by the IR's of the space group. The expansion coefficients $\{\eta_i\}$ depend on the pressure P and the temperature T, and constitute the order parameters. Sometimes one views the set $\{\eta_1, \ldots, \eta_n\}$ as being made up of the components of a vector order parameter $\boldsymbol{\eta}$, transforming according to the IR Γ (remember the expansion coefficients have the same transformation properties as the basis functions). It is customary to write

$$\eta_i = \eta\gamma_i \quad \text{with} \quad \sum_i \gamma_i^2 = 1 \; . \tag{6.8}$$

With this definition, $\{\gamma_i\}$ describes the symmetry of the ordered state, while the scale factor η is a measure of the degree of order. In a second-order transition, η is zero above T_c; when the temperature is lowered below T_c, it gradually increases continuously from zero and eventually saturates.

In terms of $\{\eta_i\}$, the free-energy expansion has the form [6.2]

$$F(T, P, \eta) = F_0 + \eta^2 A(P, T) + \eta^3 \sum_{\alpha} B_{\alpha}(P, T) f_{\alpha}^{(3)}(\gamma_i)$$

$$+ \eta^4 \sum_{\alpha} C_{\alpha}(P, T) f_{\alpha}^{(4)}(\gamma_i) + \ldots \; , \tag{6.9}$$

where $f_\alpha^{(3)}, f_\alpha^{(4)}, \ldots$ are polynomials of the third, fourth . . . orders formed from the quantities γ_i, which are invariant under the action of G. The index α is summed over all the distinct invariants which can be formed corresponding to a given order. Corresponding to the second order there is only one invariant, and it is a sum of squares. After the normalization (6.8), the second order term simplifies to what is shown in (6.9). The first order term vanishes, as it is not possible to construct a first-order invariant for a nontrivial Γ.

6.2.2 Liquid–Solid Transition

Let us now consider the liquid–solid transition. In this case, η_i's must be chosen such that they transform according to one of the IR's of E(3). The nature of the latter has been discussed, for example, by *Elliott* and *Dawber* [6.3] who point out that corresponding to every IR, there are an infinite number of basis functions of the form $u_m \exp(i\mathbf{k} \cdot \mathbf{r})$. Here m labels the IR's of rotations about a fixed axis, and can take on values $0, \pm 1, \pm 2, \ldots$, while $\exp(i\mathbf{k} \cdot \mathbf{r})$ comes from translational symmetry. Corresponding to a given IR, m is fixed while $\mathbf{k}$ varies such that $|\mathbf{k}|$ remains constant.

Turning now to the Landau theory, one first makes the simplifying assumption $m = 0$, so that $\phi_i \sim \exp(i\mathbf{k}_i \cdot \mathbf{r})$, with $|\mathbf{k}_i|$ constant. The latter implies that $\delta\varrho(\mathbf{r})$ is dominated by waves of one particular wavelength, in turn triggering (below the freezing point) translational order controlled by that wavelength scale. The i summation would now range over all $\mathbf{k}_i$ with $|\mathbf{k}_i| = $ constant. However, in practice, the i-summation is restricted to a discrete collection of wavevectors referred to as a *star*, one example of which is shown in Fig. 6.2. The geometry of the star encodes the orientational symmetry of the ordered state. For η_i one writes $\varrho(\mathbf{k}_i)$ which is in general complex. Since $\delta\varrho$ is real, one has

$$\delta\varrho(\mathbf{r}) = \sum_i [\varrho(\mathbf{k}_i)\exp(i\mathbf{k}_i \cdot \mathbf{r}) + \text{c.c.}] \tag{6.10a}$$

$$= \sum_i \varrho(\mathbf{k}_i)\exp(i\mathbf{k}_i \cdot \mathbf{r}) \ . \tag{6.10b}$$

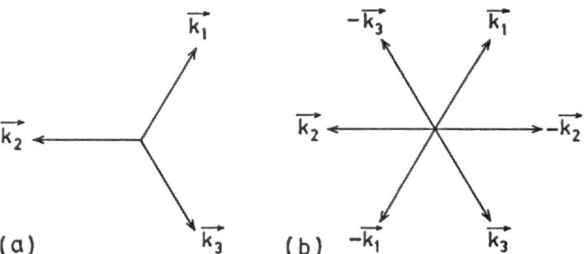

Fig. 6.2. (a) the star for 2D-hexagonal ordering if (6.10a) is used; (b) the star appropriate to (6.10b)

In (6.10a), the sum is over a star of the type shown in Fig. 6.2a, while in (6.10b) it is over the set $\{\pm k_i\}$ as in Fig. 6.2b. Since $\delta\varrho$ is real, one has

$$\varrho(-k_i) = \varrho^*(k_i) \ . \tag{6.11}$$

In the notation of (6.10b),

$$
\begin{aligned}
F = F_0 &+ \int dk' \, A(k', P, T)\varrho(k')\varrho(-k') \\
&+ B(|k|, P, T) \sum_{ijl} \varrho(k_i)\varrho(k_j)\varrho(k_l)\delta(k_i + k_j + k_l) \\
&+ \sum_{ijlm} C(\hat{k}_i \cdot \hat{k}_j, \hat{k}_l \cdot \hat{k}_m, P, T)\varrho(k_i)\varrho(k_j)\varrho(k_l)\varrho(k_m) \\
&\times \delta(k_i + k_j + k_l + k_m) + \ldots \ .
\end{aligned}
\tag{6.12}
$$

Here $\hat{k}_i$ denotes a unit vector in the direction of k_i and $|k| = |k_i|$. The magnitude $|k|$ is fixed by the minimum of the quadratic coefficient $A(k')$.

The rotational and translational symmetries of the liquid state are fully implemented in (6.12). In fact, the δ-functions arise from translational invariance. To see this, imagine the liquid to be translated by an arbitrary vector t. The $\varrho(k_i)$'s then transform as

$$\varrho(k_i) \rightarrow \varrho(k_i)\exp(ik_i \cdot t).$$

However, F must remain invariant, which implies that each term in the expansion must be unchanged. With respect to a term such as $\varrho(k_i)\varrho(k_j)\varrho(k_l)$, for example, this implies that

$$\exp[i(k_i + k_j + k_l) \cdot t] = 1 \ .$$

This is possible only if

$$k_i + k_j + k_l = 0 \ , \tag{6.13}$$

which is the restriction imposed by the δ-function. Thus, if a particular star does not have wavevectors capable of satisfying the triangle rule (6.13), then the third order term is absent.

6.2.3 BCC Versus Icosahedral Ordering

We now examine how an expansion such as (6.12) can be used to understand the occurrence of various structures. We will illustrate by considering two cases which have attracted recent attention, namely the bcc and the icosahedral structures [6.4–8]. To focus attention on the basic issue, we will use a simplified expansion due to *Mermin* and *Troian* [6.7] instead of the full-fledged one in (6.12).

Consider the expansion

$$F = A\eta^2 + B\eta^3 + C\eta^4 \ . \tag{6.14}$$

Introducing

$$\eta = -(B/C)\psi \ ,$$

substituting in (6.14) and rescaling F, we obtain

$$f = t\psi^2 - \psi^3 + \psi^4 \ . \tag{6.15}$$

Here t is proportional to $(T - T_c)$. Expansion (6.15) shows that the transition is of first order. By choosing ψ appropriate to various ordered states like bcc, 2D-hexagonal, 3D-hexagonal, icosahedral etc., one can compute the energies f associated with these various structures. Of these, Nature would prefer that for which f is the lowest. From (6.15) it is clear that the third order term would play a crucial role in this selection; in other words, the structure for which ψ^3 is largest, would win. Using such an approach and by considering appropriate stars, *Alexander* and *McTague* [6.5] showed that

$$F(\text{bcc}) < F(\text{3D–hex.}) < F(\text{2D–hex.}) < F(\text{ico.}) \ ,$$

and that bcc crystalline order is thus favoured over all others.

Alexander and McTague did their work long before the discovery of quasicrystals, their objective being to understand the wide prevalence of bcc structures (especially at high temperatures), among the elements of the periodic table. They dismissed the icosahedral structure both on account of its high energy and on account of the impossibility of having translational periodicity. It was not known then that icosahedral orientational symmetry could coexist with quasiperiodicity, although it cannot do so with strict periodicity. When quasi-crystals were finally discovered in 1984, many attempts were made to reconcile the theory with their existence. *Bak* [6.6] considered a state made up of vectors constituting the edges of a regular icosahedron as shown in Fig. 6.3a. In the free energy expansion, Bak included the fifth order term. The delta function constraint requires that the contributing terms must form a regular pentagon

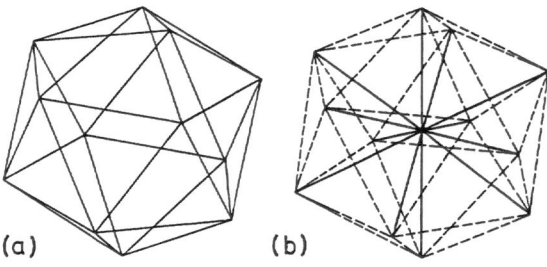

(a)   (b)

Fig. 6.3a, b. Stars for icosahedral ordering. One choice is that in (**a**) where the vectors are organized to form a (regular) icosahedron. (**b**) represents another possible choice consisting of vectors joining the centre to the vertices

which is possible for the star considered. The issue then reduces to a comparison of the $(F_3 + F_5)$ terms for the icosahedral structure with the F_3 term for the bcc structure. Making such a comparison, Bak showed that the icosahedral structure could in fact be favoured over the bcc under certain conditions.

Mermin and Troian [6.7] introduced two stars corresponding to those in Figs. 6.3a and 6.3b. The magnitudes of the wavevectors of the two stars being in the ratio $1:1.0515$, they argued that Nature would exploit both types of density waves in minimizing the free energy. Instead of (6.15), one now has [6.7]

$$f = t\psi^2 - \psi^3 + \psi^4 + (t'\phi^2 - \phi\psi^2) , \quad t' > 0 ,$$

ϕ representing the additional order parameter. Carrying out first the minimization with respect to ϕ for a fixed ψ, one obtains

$$f = t\psi^2 - \psi^3 + [1 - (1/4t')]\psi^4 .$$

Thus the additional order parameter effectively modifies the fourth order term, permitting the stabilization of the icosahedral structure in a certain temperature range (i.e., a range of t). Jaric [6.4] has pointed out that the Mermin–Troian model is equivalent to choosing the coefficient C in (6.12) in a particular way.

Kalugin et al. [6.8] also use the two stars in Fig. 6.3, but incorporate their effects by writing the density as

$$\delta\varrho(r) = \sum_i \varrho(k_i)\exp(ik_i \cdot r) + \sum_j \varrho(k_j)\exp(ik_j \cdot r) , \tag{6.16}$$

where the two sums are over the two stars referred to. The focus now shifts back to the third-order term as in the work of Alexander and McTague [6.5], and once again it is demonstrated that, under certain conditions, the icosahedral structure becomes more favourable than the bcc.

In passing we observe that, in general, the expansion (6.10) could be enlarged to include vectors resulting from various combinations of the basic set. The expansion in (6.16) offers a simple illustration. For a crystal, for instance, one similarly writes

$$\varrho(r) = \sum_{G \in \Sigma^*} \varrho(G)\exp(iG \cdot r) , \tag{6.17}$$

where the sum runs over all the vectors of the reciprocal lattice Σ^*, instead of just over vectors of a fixed length belonging to Σ^*. An expansion such as (6.17) essentially implies the inclusion of higher order terms in the Landau expansion.

6.3 Orientational Ordering

In the two examples just discussed, orientational order is automatically linked to translational order via the point group symmetry of the star. On the other hand,

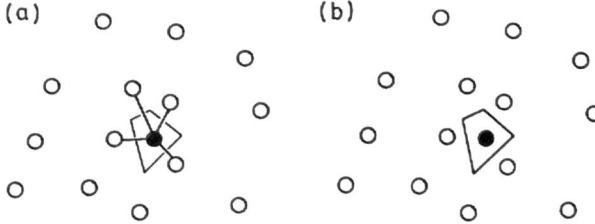

Fig. 6.4. (a) shows an assembly of atoms, and the nearest neighbour bonds associated with one of them. The bonds may be identified by constructing the Voronoi cell as in (b)

we recall that states of matter with orientational order alone are possible. A simple illustration is provided in Fig. 2.13 where every bond is parallel to one of seven basic directions. Thus, in this pattern, while bond orientations are correlated over infinite distances, the positions of the vertices are not.

To construct a Landau theory for bond orientational order, we must first have a function analogous to $\varrho(r)$. Consider the bond cluster in Fig. 6.4a. To identify the bonds surrounding the atom at r, one could construct the Voronoi or the Wigner–Seitz cell around the atom in question, as illustrated in Fig. 6.4b. With this information, one now defines $\varrho(\Omega, r)$ as the density of bonds radiating into unit solid angle around Ω. In a liquid, $\varrho(\Omega, r)$ will average to a constant, i.e., $\langle \varrho(\Omega, r) \rangle = \varrho_0$ (say).

In the orientationally-ordered state we can therefore write

$$\varrho(\Omega, r) = \varrho_0 + \delta\varrho(\Omega, r) \ . \tag{6.18}$$

As in (6.10) we expand $\delta\varrho$ in terms of suitable basis functions, which we now choose as the spherical harmonics $Y_{lm}(\Omega)$. Thus,

$$\delta\varrho(\Omega, r) = \sum_{l=0}^{\infty} \sum_{m=-l}^{+l} Q_{lm}(r) \, Y_{lm}^*(\Omega) \ . \tag{6.19}$$

Although in (6.19) l is summed over all possible values, in practice, one particular value dominates, just as one value of $|k|$ dominates in (6.10). For instance, $l = 2$ favours nematic ordering, $l = 4$ supports cubic ordering, $l = 6$ promotes icosahedral *orientational* order, and so on. It is worth stressing here that icosahedral orientational order means long-range bond-orientational order alone. The ordering discussed in Sect. 6.2.3 refers to long-range quasiperiodic translational order *plus* icosahedral orientational order. Parenthetically, we also note that the Q_{lm}'s above are a specific realization of the Q_n's introduced formally in (2.5).

Returning to (6.19) we may write, bearing in mind the comment about l,

$$\delta\varrho(\Omega, r) = \sum_{m} Q_{lm}(r) \, Y_{lm}^*(\Omega) \ . \tag{6.20}$$

The expansion coefficients are the bond-orientation order parameters, and the Landau expansion must be written using invariants formed from the Q_{lm}'s.

As an example, consider icosahedral ordering. In this case the first few invariants are [6.9]:

$$f^{(2)} = \sum_{m=-6}^{+6} (-1)^m Q_{6,m} Q_{6,-m} , \tag{6.21a}$$

$$f^{(3)} = \begin{bmatrix} 6 & 6 & 6 \\ m_1 & m_2 & m_3 \end{bmatrix} Q_{6m_1} Q_{6m_2} Q_{6m_3} , \tag{6.21b}$$

$$f_1^{(4)} = (f^{(2)})^2 , \tag{6.21c}$$

$$f_2^{(4)} = (-1)^n \begin{bmatrix} 6 & 6 & 6 \\ m_1 & m_2 & n \end{bmatrix} \begin{bmatrix} 6 & 6 & 6 \\ -n & m_3 & m_4 \end{bmatrix} \times Q_{6m_1} Q_{6m_2} Q_{6m_3} Q_{6m_4} , \tag{6.21d}$$

$$f_3^{(4)} = (-1)^{(n_1+n_2+n_3+n_4)} \begin{bmatrix} 6 & 6 & 6 \\ -n_1 & m_2 & n_2 \end{bmatrix} \begin{bmatrix} 6 & 6 & 6 \\ -n_2 & m_2 & n_3 \end{bmatrix}$$
$$\times \begin{bmatrix} 6 & 6 & 6 \\ -n_3 & m_3 & n_4 \end{bmatrix} \begin{bmatrix} 6 & 6 & 6 \\ -n_4 & m_4 & n_1 \end{bmatrix} Q_{6m_1} Q_{6m_2} Q_{6m_3} Q_{6m_4} . \tag{6.21e}$$

In the above, $\begin{bmatrix} 6 & 6 & 6 \\ m_1 & m_2 & m_3 \end{bmatrix}$ is the Wigner 3-j symbol [6.3]. It is nonvanishing only if $(m_1 + m_2 + m_3) = 0$. Summation over repeated indices is also assumed.

In terms of the invariants in (6.21), we can write

$$F = F_0 + A f^{(2)} + B f^{(3)} + \sum_{\alpha=1}^{3} C_\alpha f_\alpha^{(4)} . \tag{6.22}$$

Landau expansions of the above type have been employed in several studies to explore various aspects of orientational ordering [6.10–12].

6.4 Orientational Order Versus Translational Order

Occasionally there are several competing order parameters, each trying to establish its own brand of order. For example, one sometimes finds a sequence of structural phase transitions driven by soft modes [6.13]. This sequence is the outcome of a competition between various soft modes, each of which finally has its say in some temperature region. Similarly, a competition between orientational and translational order parameters has also been considered [6.11]. We will illustrate this by considering the work of *Jaric* [6.4].

The basic idea is that there are two versions of ordering possible: (i) pure orientational ordering of the icosahedral type described by the parameters $Q_{6m}(r)$, and (ii) quasicrystalline ordering with icosahedral point group symmetry, governed by $\varrho(k_i)$'s as discussed in Sect. 6.2.3. To analyze which of these two options Nature would prefer, one writes

$$F = F(\text{orientational}) + F(\text{quasiperiodic}) + F(\text{interaction}) \ ,$$

where the first two terms on the r.h.s correspond to Landau expansions already discussed, while the third term takes care (in a rotationally invariant way) of the interaction between the two sets of competing order parameters.

Let T_o and T_t denote the orientational and (quasiperiodic) translational ordering temperatures, in the absence of F (interaction). Jaric finds that if $T_t > T_o$, translational ordering prevails. Of course, orientational ordering is also present. However, if $T_o > T_t$, there is an intermediate phase with long range orientational order alone and no translational order. At sufficiently low temperatures, translational order also appears.

6.5 Landau Theory and Amorphous Structures

In Chap. 5, we have seen how amorphous structures can be derived from polytopes by a projection technique. Nelson and coworkers [6.14–16] also have tried to describe the DRP structure using the $\{3, 3, 5\}$ polytope as the reference template, but in the Landau-theory language.

In this approach, the local density in the DRP structure is described in terms of the order parameter applicable to the reference polytope. Of course, unlike the regular polytope, the local order parameter in the DRP structure would vary in space, vanishing completely in a region where there is a defect, i.e., a Frank–Kasper line. This behaviour of the order parameter is analogous to the case of a type II superconductor placed in a magnetic field (for further remarks on this analogy, see [6.17] and also Sect. 6.6.5). The spirit of Nelson's approach may be appreciated by considering the density expansion

$$\varrho(r) = \sum_G \varrho(G) \exp(iG \cdot r) \tag{6.23}$$

for a crystal, where G denotes a reciprocal lattice vector. By Fourier inversion, we find the order parameters to be given by

$$\varrho(G) = \frac{1}{V} \int dr \, \varrho(r) \exp(-iG \cdot r) \ . \tag{6.24}$$

In a perfect crystal, $\varrho(G)$ is independent of r, but in a defective crystal, this would not be true, and the order parameter would be a function of r. Idealizing the liquid near the freezing point as a crystal with (perhaps a substantial density of) defects, one can introduce a local order parameter defined by

$$\varrho(r; G) = \frac{1}{\Delta V_r} \int_{\Delta V_r} dr' \, \varrho(r') \exp(-iG \cdot r') \; , \tag{6.25}$$

where the integration is over a finite volume ΔV_r enclosing r. This volume ΔV_r is chosen to be larger than an atomic volume but smaller than a macroscopic one.

Nelson's philosophy is similar. Consider, for illustration, the curved space S^1, and let us suppose it is decorated with p equally-spaced atoms. The density distribution $\varrho(\theta)$ associated with this regular pattern can be formally expanded as [compare with (6.23)],

$$\varrho(\theta) = \sum_{m = 0, \pm p, \pm 2p, \ldots} \psi_m \exp(im\theta) \; . \tag{6.26}$$

The order parameter ψ_m is then given by

$$\psi_m = \frac{1}{2\pi} \int_0^{2\pi} d\theta \, \varrho(\theta) \exp(-im\theta) \; . \tag{6.27}$$

If $\varrho(\theta)$ does not correspond to a regular tiling, but has only short-range order, we can define a local order-parameter $\psi_m(\theta_0)$ in the spirit of (6.25) by writing

$$\psi_m(\theta_0) = \frac{1}{\Delta\theta} \int_{\Delta\theta} d\theta \, \varrho(\theta) \exp(-im\theta) \; , \tag{6.28}$$

where $\Delta\theta$ is a small angular interval enclosing θ_0.

Let us now consider the random chain in Fig. 6.5a [6.16]. Suppose S^1 is placed tangential to the chain at x. Project stereographically all the atoms of the chain, in a small region around x, onto S^1. The projected density $\varrho(\theta, x)$ then has a bounded range in θ. Also, at the point of tangency, the projected density must

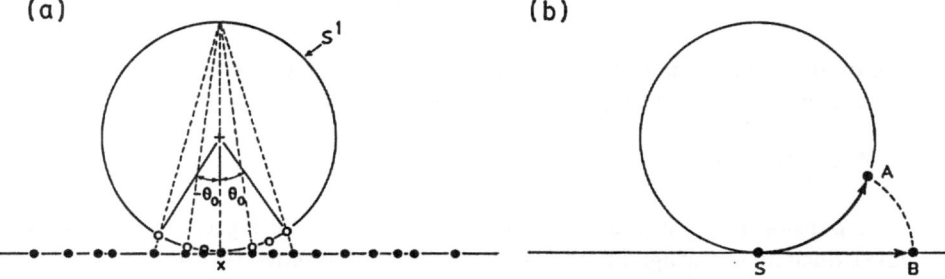

Fig. 6.5. (a) shows an amorphous chain. To define the order parameter at a point x, one places S^1 tangential to the chain, projects the atom positions to construct $\varrho(\theta, x)$ and then defines $\psi_m(x)$ as in (6.31). (b) illustrates schematically how ψ_m at S may be transported to B for defining the derivative of $\psi_m(x)$ appropriately

be equal to the density along the chain at x, i.e.,

$$\varrho(\theta = 0, x) = \varrho(x) \ . \tag{6.29}$$

In the spirit of (6.26–28), we now write

$$\varrho(\theta, x) = \sum_m \psi_m(x) \exp(im\theta) \ , \quad \text{with} \tag{6.30}$$

$$\psi_m(x) = \frac{1}{2\theta_0} \int_{-\theta_0}^{+\theta_0} d\theta \ \varrho(\theta, x) \exp(-im\theta) \tag{6.31}$$

(Fig. 6.5a). The quantity $\psi_m(x)$ describes the local order at the point x in the chain, in terms of the idealized quantity ψ_m relevant to perfect ordering in S^1. The Landau expansion is now constructed using the local order parameter $\psi_m(x)$. But before writing down this expansion, there is one more detail to consider.

Since ψ_m depends on x, the gradient term in the Landau expansion cannot be ignored, recall (3.9). Equation (6.31) tells us how ψ_m can be defined at any given x, but there is no rule as yet for comparing the values of ψ_m at adjacent points such as $S(=x)$ and $B(=x + dx)$ in Fig. 6.5b. Such a comparison is required for defining the derivative. The necessary comparison is made by resorting to a technique known as *parallel transport*. (See Appendix B.) Restricting ourselves here to a descriptive statement, comparing the values of ψ at S and B in Fig. 6.5a essentially involves first transporting the order parameter from S to a suitable point A on S^1, and then rolling S^1 along the chain so that A falls upon B (Fig. 6.5b). The true derivative involves a comparison of the order parameter at S and A respectively; it turns out that we must compensate for the curvature of S^1 by replacing $d\psi_m/dx$ with $(d/dx - im\kappa)\psi_m$, where κ^{-1} is the radius of the circle S^1. The Landau expansion associated with $\psi_m(x)$ thus reads

$$F_m(x) = A_m |\psi_m(x)|^2 + O(|\psi_m(x)|^4) + \frac{1}{2} K_m \left| \left(\frac{d}{dx} - im\kappa \right) \psi_m(x) \right|^2 \ . \tag{6.32}$$

The quantity $F_m(x)$ describes the free energy density at x. The total energy $\mathscr{F}$ associated with the field $\psi_m(x)$ is then given by

$$\mathscr{F} = \int dx \ F_m(x) \ . \tag{6.33}$$

Expansion (6.32) is well suited for studying an ordering characterized by $\psi_m(x)$. Such an ordering, if it occurred, would be dominated by one m value. On the other hand, in the disordered phase and at a temperature just above the (hypothetical) transition temperature, all m values would make a contribution. Owing to the smallness of the order parameter, the fourth order term would be negligible. Thus a quadratic approximation to F_m would be adequate, and in terms of the Fourier coefficients

$$\psi_m(q) = \int_{-\infty}^{\infty} dx \, \psi_m(x) \exp(-iqx) \, , \tag{6.34}$$

one has

$$\mathscr{F} = \tfrac{1}{2} \sum_m \sum_q |\psi_m(q)|^2 \, [K_m(q-m\kappa)^2 + A_m] \, . \tag{6.35}$$

Now the probability of occurrence of a particular order parameter fluctuation leading to an energy $\mathscr{F}$ is given by $\exp(-\mathscr{F}/k_B T)$. Using the Gaussian approximation (6.35) for $\mathscr{F}$, it follows [6.2, 15] that

$$\langle |\psi_m(q)|^2 \rangle = \frac{k_B T}{K_m(q-m\kappa)^2 + A_m} \, . \tag{6.36}$$

The structure factor $S(q)$, which is the Fourier transform of $G(r)$ defined in (2.3), is then given by

$$S(q) = \langle |\varrho(q)|^2 \rangle = (\text{constant}) \sum_m \langle |\psi_m(q)|^2 \rangle \, . \tag{6.37}$$

The structure factor is accessible from a diffraction experiment and is therefore a useful quantity to calculate.

The generalization of the above discussion to the DRP is straightforward [6.14–16]. For this purpose, Nelson and coworkers introduce a local order parameter $Q_{n,m_1 m_2}(r)$ which is a generalization of $Q_{lm}(r)$ defined earlier. This new quantity arises via the expansion of the projected density $\varrho(\hat{u})$ on S^3, i.e.,

$$Q_{n,\,m_1 m_2}(r) = \int_{\Delta V} d\Omega_{\hat{u}} \, \varrho(\hat{u}) \, Y^*_{n,\,m_1 m_2}(\hat{u}) \tag{6.38}$$

(compare with 6.31). Here $\hat{u}$ is a unit four vector denoting a position on S^3, and $Y_{n,m_1 m_2}$ is a hyperspherical harmonic. Proceeding as in the 1D illustration, one finally obtains [6.15]

$$\varrho(q) = \sum_{n=12,\,20,\,24,\,\ldots} \frac{n+1}{2\pi^2} \langle |\sum_m Q_{n,mm}(q)|^2 \rangle \, . \tag{6.39}$$

The restriction on n arises from the symmetry of $\{3, 3, 5\}$, which is the source of the order parameter.

Figure 6.6 shows computed and measured structure factors for amorphous iron. It is noteworthy that the peak positions are derived from the geometry of the polytope. The adjustable parameters of the model are the counterparts of K_m and A_m in (6.36).

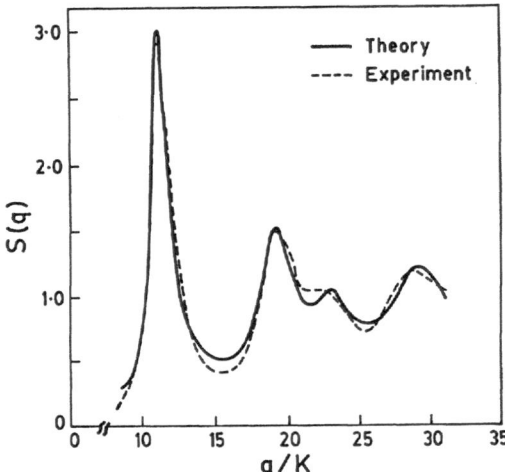

Fig. 6.6. Calculated structure factor for amorphous iron compared with experiment (After [6.16])

6.6 Landau Theory and Liquid Crystals

The Landau theory has been applied extensively to liquid crystals [6.18, 19]. Here we will illustrate such applications by considering the cases of the nematic and the smectic A phases.

6.6.1 Liquid-to-Nematic Transition

Recall first that in an ideal nematic, the molecules are (on the average) aligned parallel to one common direction (Fig. 2.8). Let $\hat{v}^{(\alpha)}$ be a unit vector along the axis of the αth molecule. This vector describes the orientation of the molecule. Since liquid crystals possess a centre of symmetry, the average of $\hat{v}^{(\alpha)}$ vanishes. By itself therefore, it is not suitable as an order parameter. A more appropriate candidate is the second rank, traceless tensor

$$Q_{ij} = \frac{1}{N} \sum_{\alpha=1}^{N} (v_i^{(\alpha)} v_j^{(\alpha)} - \tfrac{1}{3}\delta_{ij}) \ . \tag{6.40a}$$

Instead of the $\{\hat{v}^{(\alpha)}\}$ one could also use the *director* $\hat{n}$ in the definition of Q_{ij}, $\hat{n}$ denoting the *average* preferred direction of the molecules. In terms of $\hat{n}$,

$$Q_{ij} = Q(n_i n_j - \tfrac{1}{3}\delta_{ij}) \ , \tag{6.40b}$$

where the scalar quantity Q is the analogue of η in (6.8) and is a measure of the alignment of the molecules. It can be shown that the order parameters defined by (6.40) have the same transformation properties as Y_{2m}, which is expected since the liquid to nematic transition involves orientational ordering [see remarks after (6.19)]. The Landau expansion for the transition is [6.20]

$$F = F_0 + A(T)\mathrm{Tr}\{\underset{\sim}{Q}^2\} + B\,\mathrm{Tr}\{\underset{\sim}{Q}^3\} + \mathrm{O}(\underset{\sim}{Q}^4)\ . \tag{6.41}$$

A third order term is included since the transition is of first order.

The nematic can support both point as well as line defects. Let us here consider the line defects. A homotopic analysis [6.21] yields $\Pi_1 = \mathbb{Z}_2$, the group of two elements. (See also the final section of Appendix E.) Thus one expects only one class of nontrivial line defects. Such defects are known as disclinations, and were known long before homotopic classification came into existence. Figure 6.7 shows some examples of disclinations in nematics. The closure misfit (Sect. 4.4.2) is measured as illustrated in Fig. 6.8, and can take the values $2\pi S$ where S is either an integer or a half-integer. This might suggest there is a contradiction with homotopic analysis; in fact, there is none. As *Volovik* and *Mineev* [6.21] have pointed out, all disclinations characterized by integer values for S belong to the *same* class as the identity element of the first homotopy group, i.e., of $\mathbb{Z}_2$. For instance, the different $S = 1, -1$ and 2 patterns in Fig. 6.7 can be transformed into each other as illustrated in Fig. 6.9. In addition they can all be wiped out, confirming the analysis of Volovik and Mineev. Thus the only nontrivial defect is the disclination corresponding to a closure misfit of $\pm\pi$. Those with misfits $\pm3\pi$, $\pm5\pi$, etc. can all be reduced to the fundamental defect, a π-disclination.

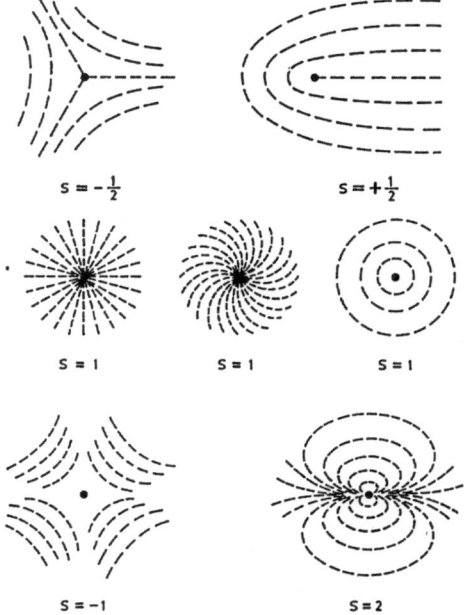

Fig. 6.7. Some of the disclinations possible in nematic liquid crystals. The director is represented by a short, headless arrow. The index S multiplied by 2π measures the closure misfit

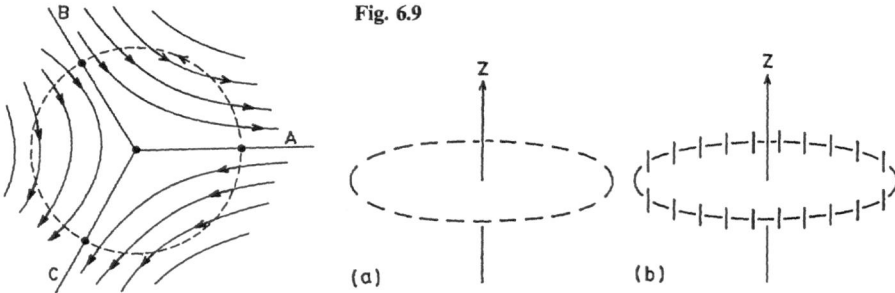

Fig. 6.9

(a)   (b)

Fig. 6.8. Scheme for measuring S. We first construct unit vectors tangential to the molecules. We then traverse the Burger's circuit ABCA, and observe the rotation of the unit vector. In the case shown it is $-\pi$, whence $S = -(1/2)$. See also Fig. 6.7

Fig. 6.9a, b. This figure illustrates how an $S = 1$ disclination (see also Fig. 6.8) can be wiped out by aligning all the directors parallel to the z axis. (This is sometimes called "escape into the third dimension".) An $S = 1/2$ disclination cannot be wiped out in a similar manner by rotating all directors by the *same* amount

6.6.2 Deformation Energy

The ordered state predicted by (6.41) is spatially uniform. One can impose on this a distortion described by a variable director $\hat{n}(r)$. The three important types of distortion are the splay, twist and bend illustrated in Fig. 6.10. Their contribution to the elastic energy is [6.20]

$$F_{el} = \tfrac{1}{2}K_1(\nabla \cdot \hat{n})^2 + \tfrac{1}{2}K_2(\hat{n} \cdot \nabla \times \hat{n})^2 + \tfrac{1}{2}K_3(\hat{n} \times (\nabla \times \hat{n}))^2 \ , \tag{6.42}$$

(SPLAY)   (TWIST)   (BEND)

where K_1, K_2 and K_3 are elastic constants associated with the three basic types

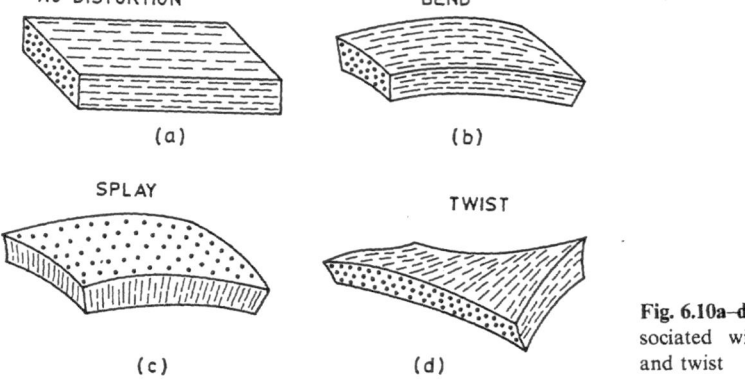

NO DISTORTION   BEND

(a)   (b)

SPLAY   TWIST

(c)   (d)

Fig. 6.10a–d. Distortions associated with bend, splay and twist

of deformation. Expression (6.42) is the specific form of the gradient energy (introduced in Sect. 3.1.6) for the case of the nematic.

6.6.3 Nematic-to-Smectic A (NA) Transition

The transition to smectic A occurs by the breaking of translational symmetry along one of the three dimensions (recall Table 3.2). To discuss this transition we therefore proceed as in Sect. 6.2, writing [6.22]

$$\varrho(r) = \frac{1}{\sqrt{2}} \left[\varrho(k) \exp(ikz) + \text{c.c.} \right] , \tag{6.43}$$

where $k = (2\pi/d)$, d being the layer spacing. Following the convention in this field, let us write ψ in place of $\varrho(k)$. In terms of ψ, the free energy expansion is

$$F = F_0 + A(T)|\psi|^2 + C|\psi|^4 . \tag{6.44}$$

Next we must include the energy associated with spatial variations of the order parameter. This is given by a term of the form (with $\partial_i \equiv \partial/\partial x^i$)

$$\tfrac{1}{2} \mu_{ij} (\partial_i \psi^*)(\partial_j \psi) , \tag{6.45}$$

the tensor μ_{ij} suitably taking care of the anisotropy represented by the layering. Expression (6.45) does not however allow for interplay between the spatial variations of ψ and those of the director. When this is done, the gradient term becomes [6.22]

$$F_{\text{grad}} = \frac{1}{2} \mu_{ij} \left[\left(\partial_i - \frac{2\pi i}{d} \delta n_i \right) \psi^* \right] \left[\left(\partial_j + \frac{2\pi i}{d} \delta n_j \right) \psi \right] , \tag{6.46}$$

where we have written $\hat{n} = \hat{n}_0 + \delta n$, $\hat{n}_0$ being the undisturbed orientation of the director. Once we permit the director to fluctuate, then we must also allow for an elastic deformation energy as in (6.42). Thus the net free energy density for describing the NA transition is

$$F = F_0 + A(T)|\psi|^2 + C|\psi|^4 + F_{\text{grad}} + F_{\text{el}} , \tag{6.47}$$

where F_{el} is given by (6.42). This is referred to as the de Gennes model [6.22].

6.6.4 Defects in Smectic A

On account of the translational ordering in the smectic A phase, one should expect dislocations in smectic A as illustrated in Fig. 6.11a. A closer examination reveals, see e.g. [6.23], that the dislocation is in fact made up of two elementary disclinations as shown in Fig. 6.11b. Parenthetically we remind the reader that periodicity in one dimension is strictly not possible in the sense of crystalline

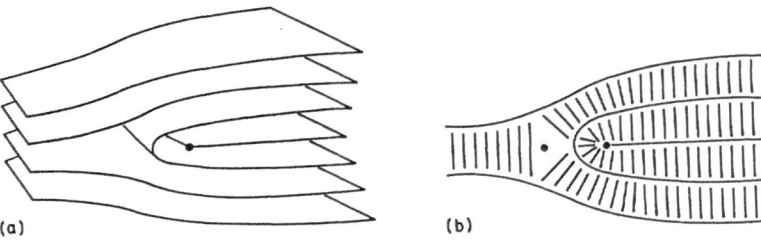

Fig. 6.11. (a) shows a dislocation in the smectic A phase in terms of the layers associated with translational order. A closer look as in (b) reveals that the dislocation is made up of two disclinations

order, as pointed out in Sect. 3.6. However, there is a stacking of layers, and to that extent one can certainly talk of dislocations and Burgers vectors.

6.6.5 Analogies to the Superconductor

The smectic A has many parallels to the superconductor. Recall first that there are two types of superconductors, namely, types I and II, distinguished by differences in their magnetic properties. Type I materials show the classic Meissner effect, i.e., total exclusion of the applied field H, provided $H < H_c$, the critical field. Above H_c, the material becomes normal. Type II materials, on the other hand, exhibit the Meissner effect up to a field H_{c_1}, permit partial penetration up to a field H_{c_2}, and then become normal. When $H_{c_1} < H < H_{c_2}$, the material accommodates the applied field by becoming normal in select regions. More explicitly, the applied field is distributed as an array of flux filaments, arranged in fact on a lattice as illustrated in Fig. 6.12a. This mixture of normal and superconducting regions is sometimes referred to as the Abrikosov–Shubnikov phase.

The flux filaments are vortex singularities, and the structure of a typical vortex is illustrated in Figs. 6.12 b, c. There is a core region of dimension $\sim \xi$ (the "coherence length") where the material is normal and facilitates flux penetration. The flux itself spills over somewhat into the superconducting region, the penetration being characterized by a depth λ. Depending on whether $\lambda \ll \xi$ or $\lambda \gg \xi$, one has a type I or type II superconductor.

The Ginzburg–Landau expansion for a superconductor in a magnetic field is given by [6.24]

$$F(\mathbf{r}) = F_0 + A|\psi(\mathbf{r})|^2 + C|\psi(\mathbf{r})|^4 + \gamma \left| \left(\nabla - \frac{\mathrm{i}e^*}{c} A \right) \psi \right|^2$$

$$+ \frac{1}{8\pi\mu} \sum_{i > j} (\partial_j A_i - \partial_i A_j)^2 \ . \tag{6.48}$$

Here ψ is the (complex) order parameter, $e^* (= 2e)$ is the effective charge, and μ is

Fig. 6.12. (a) flux lattice in a type II superconductor. (b) shows a typical filament and the circulating supercurrent. The spatial variation of the magnetic field and the order parameter field are sketched in (c) and (d) respectively

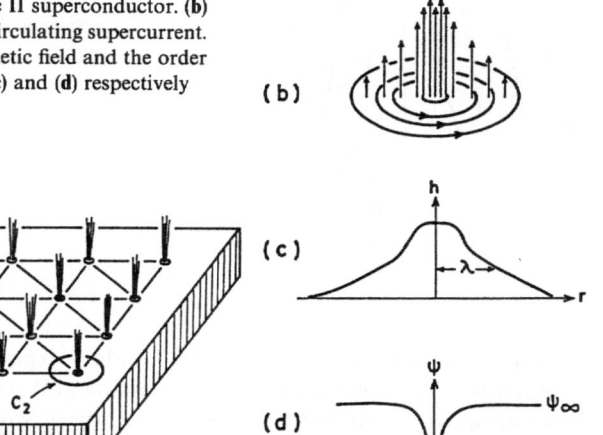

the permeability of the normal metal. The vector potential $A(r)$ is introduced to take care of the effects of the magnetic field. In the absence of the latter,

$$F(r) = F_0 + A|\psi|^2 + C|\psi|^4 + \gamma|\nabla\psi|^2 , \tag{6.49}$$

a form already familiar to us.

There are clearly parallels between (6.48) and (6.47) [6.22]. We observe, for example, that $\delta n(r)$ plays a role analogous to the vector potential $A(r)$. Indeed there are further similarities, and one can define a coherence length ξ as well as a penetration depth λ. The quantity ξ is the radius of the core of the dislocation, while λ is the thickness to which a weak twist or bend deformation applied at the surface penetrates. Based on these two quantities, one can classify smectics into types I and II, depending on the value of (λ/ξ). The phase diagrams for the two cases are as sketched in Fig. 6.13a, b. In the "Abrikosov–Shubnikov phase" which results when a weak bend or twist is applied at the free surface, one expects a network of edge dislocations as illustrated in Fig. 6.14.

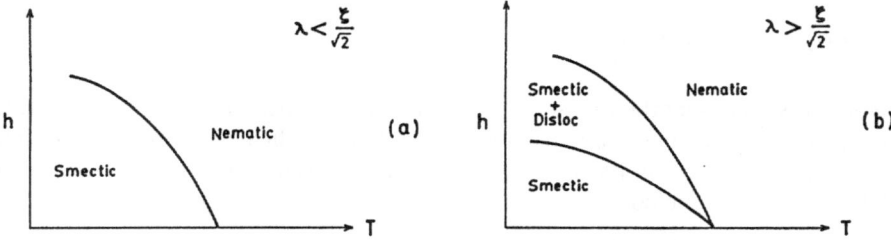

Fig. 6.13a, b. Phase diagrams for the nematic–smectic A system. (a) and (b) correspond to type I and type II behaviours respectively. h is the applied mechanical force

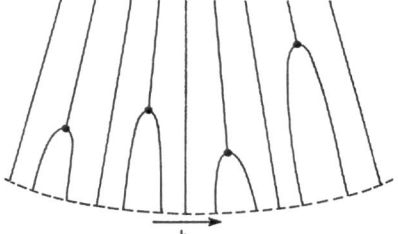

Fig. 6.14. Array of dislocations induced by a bend in the smectic A phase

The structure of the Ginzburg–Landau expansion (6.48) for the super-conductor is governed in part by a fundamental requirement known as gauge invariance (Appendix F; for an elementary discussion, see [6.17]). In the case of smectic A, however, there is no requirement of gauge invariance although the director field $\delta \hat{n}(r)$ plays a role analogous to the "gauge field" $A(r)$.

6.7 Hydrodynamics

The term hydrodynamics is used in two related but somewhat different senses in condensed matter physics. Frequently it just refers to the long wavelength behaviour (i.e., the $q \to 0$ limit) of the elementary excitations associated with the fluctuations of the order parameter. On the other hand, in the study of fluids the term hydrodynamics refers to mass, momentum and energy transport. The mathematical theory devised for fluids has been extended to condensed matter systems [6.25, 26], making due allowance for additional flows associated with the order parameter field. For example, in a ferromagnet there can be a flow of magnetization. Unified hydrodynamics will be treated in Appendix A, but here we will illustrate with an example what is meant by hydrodynamics in the restricted sense.

Consider a quasicrystal Q with density

$$\varrho(r) = \sum_{G_i} \varrho(G_i) \exp(i G_i \cdot r) \ . \tag{6.50}$$

Here the sum is over the vectors of Σ^*, the reciprocal lattice for Q. Σ^* is the lattice reciprocal to the (higher dimensional) lattice Σ from which Q is projected. Using G_i in place of k_i to emphasize that the vectors concerned belong to a reciprocal lattice, we write the complex amplitude as

$$\varrho(G_i) = |\varrho(G_i)| \exp[i\phi(G_i)] \ . \tag{6.51}$$

If Q is a perfect quasicrystal, the phases $\phi(G_i)$ are constant throughout the medium. However, fluctuations and topological defects can cause the phases to be r-dependent, recall also (6.25). If $|\varrho(G_i)|$ is constant, then the gradient or the elastic energy arises from the r-dependence of the phase angles.

To set up a theory of elastic deformations of the perfectly ordered state, we need to know how many *independent* $\phi(G_i)$'s there are. These are referred to as *hydrodynamic variables*. It has been established [6.27] that for periodic and quasiperiodic systems, the number N_R of hydrodynamic variables is equal to the number of *independent* vectors required to generate the reciprocal lattice. Recalling earlier results from Chap. 5, we see that $N_R = 4$ for a (2D) Penrose tiling, while $N_R = 6$ for a 3D icosahedral structure. These are to be compared with the numbers 2 and 3 respectively for 2D and 3D periodic lattices.

At this stage, let us specialize to the case of the 3D icosahedral quasicrystal in order to fix our ideas. By recalling Fig. 6.2b, we see that there are six independent phase variables governing this structure. These six phases can be replaced by two vectors u and w, each with three components [6.27, 28], according to

$$\phi(G_i) = G_i \cdot u + H_i \cdot w, \quad (i = 1, \ldots, 6) , \tag{6.52}$$

where the vectors $\{H_i\}$ are a suitably defined permutation of $\{G_i\}$. The components of u describe the familiar displacements, whereas w represents the relative motion of the constituent density waves. The fields u and w transform under the irreducible representations Γ_3 and Γ_3' of the icosahedral point group (Table 5.4). Based on these properties and the constraints of translational and rotational invariance, the gradient energy density or the harmonic elastic free energy for the icosahedral crystal may be written as

$$F(u, w) = F_u + F_w + F_{uw} . \tag{6.53a}$$

Here F_u is the standard elastic energy,

$$F_u = \tfrac{1}{2}\lambda u_{ii} u_{jj} + \mu u_{ij} u_{ij} , \tag{6.53b}$$

λ and μ are the Lamé constants, and

$$u_{ij} = \tfrac{1}{2}(\partial_i u_j + \partial_j u_i) , \quad \partial_i u_j \equiv \partial u_j / \partial x_i \tag{6.54}$$

is the elastic strain tensor. F_w and the coupling term F_{uw} are involved expressions [6.28]. This expression for F, together with phenomenological dissipative terms, provides a framework for the hydrodynamic description of quasicrystals. One now proceeds as in phonon physics [6.29, 30] to obtain the dispersion behaviour of the long wavelength modes associated with the fields $u(r)$ and $w(r)$. One finds [6.28] that in addition to the usual phonon modes, there are three *phason* modes. Phasons occur whenever there is an incommensuration of some sort.

6.8 Fluctuations and the Landau Theory

It is pertinent therefore to consider the extent to which the Landau–Ginzburg theory allows for fluctuations of the order parameter, and if so, to what extent [6.31]. In answering this question, we must first note that the Landau theory is "mesoscopic" rather than microscopic. The density in (6.7) is a coarse-grained one, i.e., it is obtained by averaging over a volume $L^3 \gg a^3$, where a is an atomic dimension. However, L is still smaller than the size of the sample (assumed to be macroscopic in extent). The coarse-graining effectively averages over fluctuations with wavelengths $\lesssim 2\pi L$, i.e. $k \gtrsim L^{-1}$. The coarse-grained order parameter is suited for discussing slow variations of the order parameter with wavelengths $\gtrsim 2\pi L$ i.e. $k \lesssim L^{-1}$. The gradient energy is in fact the energy associated with such slow variations.

Next comes the choice of L. In the Landau theory, $a \ll L \ll \xi$ where ξ is the correlation length (see Sect. 2.6). Also L is held fixed as $T \to T_c^+$. Apart from this restriction, the choice of L is arbitrary. Implicit in the coarse graining procedure is the assumption that all the important high frequency fluctuations have been averaged over. While this might be true for $T \gg T_c$, as $T \to T_c^+$, fluctuations will appear with wavelengths greater than $2\pi L$. One must therefore take steps to include also the fluctuations of longer and longer wavelengths which appear as one approaches T_c. In other words, care is needed when applying Landau theory close to T_c. The need for such caution becomes evident when one compares some of the predictions of the theory with experiment.

Consider the familiar problem of ferromagnetism. In the Landau theory, one discusses the ferromagnetic transition via the expansion

$$F(r) = F_0 + A M^2(r) + C M^4(r) + K |\nabla M|^2 , \qquad (6.55)$$

where $M(r)$ is the magnetization and represents the order parameter. Just below T_c, the spontaneous magnetization varies as $(T_c - T)^\beta$. The Landau theory predicts a value of 0.5 for the exponent β, whereas experiment yields values in the neighbourhood of 0.35. Similarly, for the correlation length ξ, which behaves as $(T - T_c)^{-\nu}$ as $T \to T_c^+$, Landau's theory predicts $\nu = 0.5$, while from experiments one finds that ν is in the range 0.6–0.7.

The discovery of the cure for these discrepancies is one of the exciting sagas of modern physics [6.32]. In brief, one uses a technique known as the *renormalization group* to include progressively the effects of long wavelength fluctuations as $T \to T_c$. In simple terms, it is as if one increases L in steps to accommodate the fluctuations of longer wavelengths. Correspondingly, the coefficients A and C in (6.55) become L-dependent, this dependence being described by suitable differential equations [6.31–33]. It is these equations which lead to improved values for the critical exponents.

The shortcomings of the Landau theory close to T_c can also be understood in a slightly different manner. Consider one particular pattern $M(r)$ of the order parameter field. The energy $\mathscr{F}$ associated with this pattern is given by

$$\mathscr{F}[M] = \int dr \ F(r) \ , \tag{6.56}$$

where $F(r)$ is as in (6.55). We write $\mathscr{F}[M]$ to denote that $\mathscr{F}$ is a *functional* of M, i.e., it depends on the form of $M(r)$. Bearing in mind that the probability for the occurrence of a particular pattern $M(r)$ with energy $\mathscr{F}$ is proportional to $\exp(-\mathscr{F}/k_B T)$, the partition function is given by

$$Z = \int d[M(r)] \ \exp(-\beta \mathscr{F}[M]) \ , \qquad \beta = (k_B T)^{-1} \ , \tag{6.57}$$

where $\int d[M]$ denotes a *functional* integration, i.e., an integration over all possible forms of the order parameter field. The Ginzburg–Landau theory amounts to the assumption that there is only one configuration contributing to the partition function, namely, that for which

$$\exp(-\mathscr{F}/k_B T)$$

is a maximum, i.e., where $\mathscr{F}$ as given by (6.56) assumes its minimum. Renormalization group theory does not make such drastic assumptions. By allowing a better estimate of Z, it also leads to improved values for the critical exponents.

6.9 Frustration and the Disruption of Order

Usually the perfectly ordered state is the state of minimum (free) energy, and distortions cause an increase in energy given by the gradient term. There are, however, systems in which distortions influence energy minimization, the blue phase of the cholesterics being an example.

The blue phases of a cholesteric liquid crystal appear in a narrow temperature range between the regular cholesteric and the isotropic liquid phases. One of these is the "blue fog" which appears at the high end of the blue-phase temperature range. *Sethna* et al. [6.34] are of the view that the complex structure of the blue fog is a manifestation of the clash between various contributions to the free energy.

As usual the order parameter is a real, traceless symmetric tensor Q_{ij}. The bulk free energy is given as in (6.41) by

$$F_{\text{bulk}} = A \ \text{Tr}\{Q^2\} + B \ \text{Tr}\{Q^3\} + C(\text{Tr}\{Q^2\})^2 \tag{6.58}$$

which favours a uniaxial form for Q, i.e.,

$$Q_{ij} = Q(n_i n_j - \tfrac{1}{3}\delta_{ij}) \ , \tag{6.59}$$

where $\sum_i n_i n_i = 1$. The rotationally invariant gradient energy is given to second order by

$$F_{\text{grad}} = \tfrac{1}{2} K_1 (\nabla_k Q_{ij})^2 + \tfrac{1}{2} K_2 (\nabla_k Q_{kj})^2 + \tfrac{1}{2} K_3 \varepsilon_{kil} (\nabla_k Q_{ij}) Q_{jl} \ . \tag{6.60}$$

Here K_1, K_2, K_3 are suitable "elastic" constants, and the Levi–Civita symbol ε_{kil} is defined as follows:

$\varepsilon_{kil} = 0$ if any two indices are equal
$\quad = 1$ if (k, i, l) corresponds to a cyclic order of (x, y, z)
$\quad = -1$ if (k, i, l) corresponds to a noncyclic order of (x, y, z).

In (6.60) and hereafter, the convention of summation over repeated indices is followed.

The problem is to determine Q by minimizing $F_{\text{bulk}} + F_{\text{grad}}$. The solution to this problem is not known. Whereas F_{bulk} by itself favours uniform ordering with a uniaxial form for Q, F_{grad} is minimized by a biaxial form for Q. Clearly there is a clash between the requirements of the two terms, and the material ends up with defects.

Meiboom et al. [6.35] offer some feel for the structure of the blue fog. They start with the uniaxial order parameter as favoured by the bulk terms and attempt to build structures reducing F_{grad} as much as possible. They find that locally the medium has a double twist structure (Fig. 6.15). However, local double-twist structures in different regions cannot be matched smoothly to fill

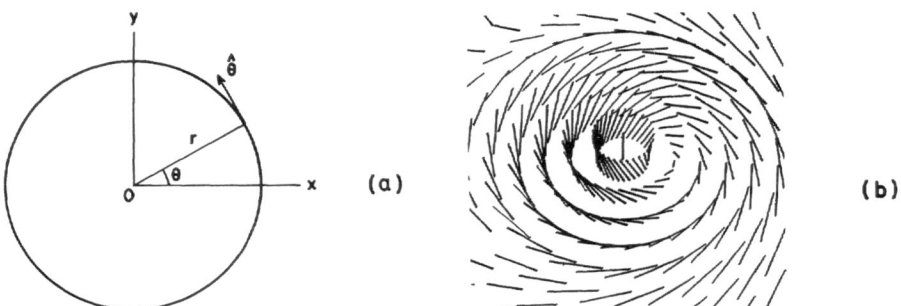

Fig. 6.15. The director field in a double-twist structure is described by

$\hat{n}(r) = \hat{z} \cos q_0 r - \hat{\theta} \sin q_0 r$

$\hat{\theta}(r) = -\hat{x} \sin \theta + \hat{y} \cos \theta$

where $\hat{\theta}$ is a unit vector as defined in (**a**). Quantities with a hat denote unit vectors. (**b**) gives a three-dimensional view of the disposition of $\hat{n}$ vectors in the plane $z = 0$. The situation in other planes is similar. As one moves along a circle of fixed radius, $\hat{n}$ twists once. As r increases, the orientation of $\hat{n}$ at any (θ, z) changes in a periodic manner. This is the second twist. The double-twist solution is valid only locally

(Euclidean, 3D) space, and it is necessary to introduce a network of defects rather like those in the Frank–Kasper structures.

The impossibility of extending (without defects) local double-twist ordering throughout the system is a form of frustration. It is noteworthy that unlike E^3, S^3 can be tiled with local double-twist structures. Thus blue fog could be regarded as a projection onto E^3 of a regular tiling of S^3 with double-twist structures [6.34].

6.10 Defect-Dominated Structures

We have seen how both translational and orientational order may be under-stood within the broad framework of the Landau theory (Sect. 6.2, 3), as also the short-range order in some amorphous structures (Sect. 6.5). It remains now to add some comments about systems with quasi-LRO.

Earlier we have pointed out (Sect. 3.6) that quasi-LRO occurs in 2D solids at low temperatures. Can an order parameter description be constructed to describe such a system? *Kosterlitz* and *Thouless* [6.36] observed that a descrip-tion based on topological defects would be more appropriate. This concept is briefly reviewed in the present section.

6.10.1 Role of Topological Defects

To understand how topological defects can control structures and mediate in phase transitions, let us consider some spatial patterns associated with the 2D XY model. Figure 6.16 shows the spin configuration one would expect at low temperatures—not quite perfect ordering, but sufficient alignment to ensure quasi-LRO (i.e., power-law decay of the spin correlation function). We have already seen in Fig. 4.1 what the topological defects in this system look like.

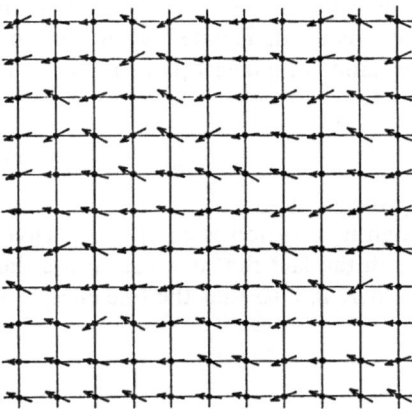

Fig. 6.16. Spin pattern in a 2D XY system at low temperatures. Perfect ordering is inhibited by fluctuations

From that figure it is clear that the disturbance due to the singularities spreads out to infinite distances. One might therefore conclude that defects cannot be present at low temperatures, for they would disrupt quasi-LRO. However, this is not quite true, as can be seen from Fig. 4.7. Here we have a closely bound defect pair whose disturbance is confined to a local region. Thus a small number of tightly bound defect pairs may occur without affecting quasi-LRO significantly.

Simple energy considerations substantiate this qualitative expectation. To see this, we start with the 2D-XY model Hamiltonian

$$\mathscr{H} = -J \sum_{\langle ij \rangle} \boldsymbol{S}_i \cdot \boldsymbol{S}_j \; , \tag{6.61}$$

where $J > 0$, and the sum $\langle ij \rangle$ is over nearest neighbour spins on a square lattice (say). One may set $|\boldsymbol{S}_i| = 1$ without loss of generality so that

$$\mathscr{H} = -J \sum_{\langle ij \rangle} \cos \theta_{ij} \; ,$$

where θ_{ij} is the angle between $\boldsymbol{S}_i$ and $\boldsymbol{S}_j$. Then

$$\mathscr{H} = -J \sum_{\langle ij \rangle} \cos (\theta_i - \theta_j) \; , \tag{6.62}$$

where θ_i is the direction of $\boldsymbol{S}_i$. Using the continuum approximation, the energy of an isolated vortex is found to be [6.36]

$$E(\text{vortex}) = E_c + \int_{a_0}^{R} dr\, 2\pi r \frac{J}{2} |\nabla \theta|^2$$

$$= E_c + \pi J \ln(R/a_0) \; . \tag{6.63}$$

Here R is the radius of the system and a_0 is a length of atomic dimensions, roughly characterizing the core of the vortex. E_c is the core energy associated with the region $r < a_0$ (put in by hand!). Since E increases (logarithmically) with the size, it would naturally be difficult to create an isolated vortex in the thermodynamic limit ($R \to \infty$). On the other hand, for a defect pair the energy is finite, being given by

$$E_{\text{pair}} \simeq 2E_c + 2\pi J \ln(r/a_0) \; , \tag{6.64}$$

where r is the separation between the two vortex lines.

Going back to the single vortex, although its creation is energy-expensive, there is an entropy advantage associated with the fact that one can locate the vortex in various ways. This entropy is $2k_B \ln(R/a_0)$, so that the free energy is

$$F = E - TS$$

$$= (\pi J - 2k_B T) \ln(R/a_0) \; , \tag{6.65}$$

implying that for $T > T_c$, where

$$T_c = \pi J/2k_B \; , \tag{6.66}$$

free vortices may indeed exist, even in the thermodynamic limit.

Based on the above arguments, the following scenario may be envisaged. At low temperatures, only vortex pairs exist. These are thermally generated and do not cause serious perturbations at large distances. Above T_c, the vortex pairs dissociate since the system can support isolated vortices on account of the free energy advantage. Once isolated vortices appear, quasi-LRO is destroyed and replaced by SRO. There is thus a transition from quasi-LRO to SRO at T_c, associated with the dissociation of defect pairs. This change is referred to as a *Kosterlitz–Thouless transition*.

Essentially similar ideas apply to the melting of a 2D solid when vortex pairs are replaced by dislocation pairs. More about this shortly.

6.10.2 Topological Order

Though conventional LRO is absent, Kosterlitz and Thouless suggest that one may differentiate between the phases on either side of the transition temperature via the concept of topological order. Briefly, topological order is an assessment of whether the topological defects are bound ("ordered") or unbound ("disordered"). To make such an assessment, one makes a large closed circuit in the system and measures some quantity related to order (the phase angle in case of the spins) all along the boundary. The total change of this quantity will be determined by the number and strength of the singular points enclosed by the path. In the high temperature phase (Fig. 6.17a) there will be a preponderance of isolated singularities. The number enclosed will be proportional to the area A of the contour C, and the average total phase change will be proportional to the length L of C. In the low temperature phase (Fig. 6.17b), the average total phase

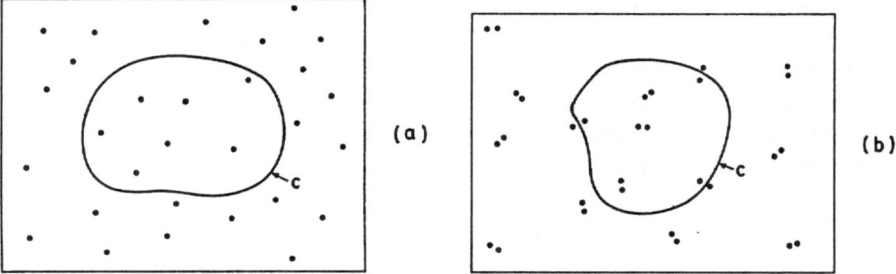

Fig. 6.17a, b. Topological order is probed by going round a simple closed contour and measuring the overall change in the phase angle. The result depends on whether the defects are free as in (a) or paired as in (b)

change will be determined by the number of pairs which cut the path, and will be proportional to $(LD)^{1/2}$ where D is the mean separation of the pairs. Based on this one can recognize a change in topological order at T_c. With respect to the correlation function, this change manifests itself as a transition from a power law to exponential decay. It is worth noting that a similar technique of assessment is used in lattice gauge theories [6.37].

6.10.3 Critical Behaviour of the Model

The problem of defect pairing and dissociation considered in Sect. 6.10.2 is conveniently discussed by considering the equivalent problem of a 2D Coulomb gas of positive and negative charges m_i with $\sum_i m_i = 0$, i.e., the gas is electrically neutral. The partition function for this system is given by [6.38]

$$Z = \mathrm{Tr}\left\{\exp\left[\pi K \sum_{i \neq j} m_i m_j \ln(r_{ij}/a_0) + \ln y \sum_j m_j^2\right]\right\} , \tag{6.67a}$$

where

$$y = \exp(-\pi^2 K/2) \tag{6.67b}$$

is the (T-dependent) fugacity and a_0 is the lattice spacing. Here K is a dimensionless parameter (analogous to $(J/k_B T)$ of the XY model). The first term in the square brackets in (6.67a) represents the Coulomb interaction in two dimensions and is the analogue of the vortex pair energy, while the second is like the core energy [in the dislocation model, $y = \exp(-E/k_B T)$].

The task now is to study the behaviour of the partition function as the temperature is increased from an initially small value, treating K and y as independent parameters. In the Kosterlitz–Thouless picture, more and more defect pairs (dipoles) with progressively larger separations are created as the temperature is raised. The problem of including the effect of these larger dipoles is handled using a renormalization group scheme as in Sect. 6.8. Focusing attention on a particular dipole of size r, the interaction between the two members constituting the dipole will be screened by (smaller) pairs lying within the range of the field. These pairs will, in turn, be screened by others in their respective fields, and so on. One thus has a scaling situation as in the case of critical fluctuations. Just as one increases L there to obtain differential equations for $A(L)$ and $C(L)$, one increases likewise the length scale in this case to obtain recursion relations for K^{-1} and y which read

$$\frac{dK^{-1}(L)}{dL} = f_1\{K^{-1}, y\} ,$$

$$\frac{dy(L)}{dL} = f_2\{K^{-1}, y\} , \tag{6.68}$$

where f_1 and f_2 are known functions.

The behaviour of the system close to T_c is best studied by considering the flows associated with (6.68). Figure 6.18 shows the relevant flow diagram where the different trajectories essentially correspond to different solutions of the coupled differential equations for different *temperatures* (T is implicit in K^{-1} and y). One follows the trajectory flows to read off the terminal values of K and y, i.e., the values they attain as $L \rightarrow \infty$. These determine the equilibrium behaviour of the system. A flow diagram like the one in Fig. 6.18 applies to all 2D systems exhibiting a Kosterlitz–Thouless transition, e.g., the 2D XY-model, 2D crystal, etc.

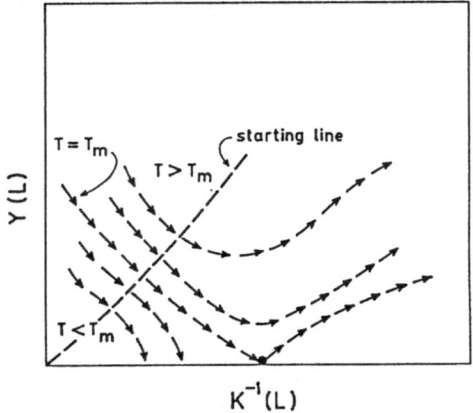

Fig. 6.18. Renormalization group flows in the Kosterlitz–Thouless theory. For explanations, see text

Our interest being in 2D solids, let us now interpret Fig. 6.18 in terms of that system. Suppose we take the dashed line in that figure as the locus of starting points. We now follow a typical trajectory for $T < T_m$ from the starting point on the line. The flow takes us to the abscissa where y is zero and K^{-1} has a finite value. Remembering that $y = \exp(-E/k_B T)$, this implies that the probability of finding a free dislocation is zero. The limiting value of K defines K_{tr}, the renormalized stiffness constant. The trajectory for $T = T_m$ is a *separatrix*. It is a dividing line, since trajectories to its right flow away. This implies that for $T > T_m$, (i) free dislocations can be found and (ii) the system loses rigidity since $K \rightarrow 0$. The behaviour of the system close to T_m may be carefully examined by focusing on the region where the separatrix curve meets the abscissa. Such an analysis yields the behaviour of the correlation function close to the transition point.

6.10.4 Disclinations and 2D Melting

In the Kosterlitz–Thouless view, a 2D solid melts at T_m owing to the dissociation of dislocation dipoles. About five years after this theory was published, *Nelson* and *Halperin* [6.39] drew attention to an important aspect of the problem not considered in the earlier treatment. They argued that a 2D dislocation was itself a composite of two disclinations as illustrated in Fig. 6.19 (see also Fig. 6.11). There must therefore be a second transition at a temperature $T_i > T_m$ where the dislocation itself breaks up into two free disclinations. Thus the isotropic liquid is essentially a 'plasma of disclinations'. When T is lowered to a value below T_i, the disclinations pair up first to form free dislocations. This pairing brings in strong orientational correlations (power law decay), but the dislocations are powerful enough to disrupt positional correlations. This is the *hexatic phase* with short-range translational order and quasi-longrange orientational order (Fig. 3.6). Below the second transition at T_m the dislocations pair up to form dislocation dipoles, elevating the positional correlations to a power law decay and the bond orientations to LRO. There has been considerable discussion in the literature on the actual existence of the hexatic phase. There appears to be some recent evidence in favour of it [6.40].

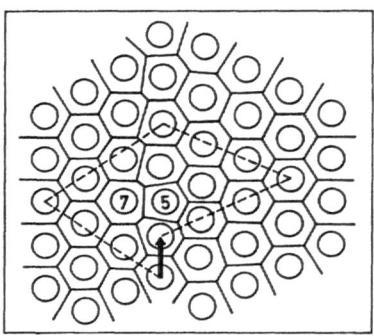

Fig. 6.19. We recall from Fig. 5.5 that a disclination in a 2D hexagonal lattice implies that the atom at the disclination site has either 5 or 7 neighbours. A 5–7 disclination pair is equivalent to a dislocation, as illustrated here

6.10.5 Landau Theory and Defect-Mediated Transitions

One might wonder whether there is any connection between the Landau and the Kosterlitz–Thouless theories. There is, in a sense. To appreciate this, let us consider the liquid-to-hexagonal lattice transition in 2D. In the Landau theory, one would first expand the density as

$$\varrho(r) = \sum_{k_i} \varrho(k_i) \exp(i k_i \cdot r) \ , \tag{6.69}$$

where k_i runs over a hexagonal star (Fig. 6.2b). Substituting the $\varrho(k_i)$'s into a standard Landau expansion leads to a perfect hexagonal lattice. But we know from the Landau–Peierls argument that a perfectly periodic lattice is not possible in 2D, which means that the conventional Landau theory is inadequate in this case. (The prediction of an ordered crystalline phase within the framework of the Landau theory is not surprising, because the latter is essentially a mean field theory which does not take cognizance of a lower critical dimensionality.) In the language of the Landau theory the source of trouble is that, in the ground state, the phase angles $\phi(\mathbf{k}_i)$ defined by

$$\varrho(\mathbf{k}_i) = |\varrho(\mathbf{k}_i)| \exp[i\phi(\mathbf{r})]$$

are $\mathbf{r}$-dependent. Transforming to a 2-component field $\mathbf{u}(\mathbf{r})$ as in (6.52) and defining a strain variable as in (6.54), the deformation energy density can be written as

$$F_u(\mathbf{r}) = \frac{\lambda}{2} u_{kk}^2(\mathbf{r}) + \mu u_{ij}(\mathbf{r}) u_{ij}(\mathbf{r}) \; , \qquad (6.70)$$

where λ and μ are the Lamé constants. This gradient energy could be added to the usual Landau expansion in terms of the $\varrho(\mathbf{k}_i)$'s as a possible cure. However, this does not straightaway predict a ground state other than the perfectly periodic one. A routine addition of a gradient energy is therefore of no help. From the work of Kosterlitz and Thouless we know now that the ground state must have bound defect pairs. Thus the Landau theory must be extended in such a manner as to accommodate topological defects and their interactions. This is done at the level of the strain variable by writing [6.39]

$$u_{ij}(\mathbf{r}) = \phi_{ij}(\mathbf{r}) + u_{ij}^{\text{sing}}(\mathbf{r}) \; , \qquad (6.71)$$

where $\phi_{ij}(\mathbf{r})$ is the smoothly varying part of the strain field while $u_{ij}^{\text{sing}}(\mathbf{r})$ is the singular part associated with the dislocations. With this decomposition, the elastic Hamiltonian is given by [6.39]

$$\mathscr{H}_E = \mathscr{H}_0 + \mathscr{H}_D \; , \quad \text{where}$$

$$\mathscr{H}_0 = \tfrac{1}{2} \int d\mathbf{r}(\lambda\phi_{kk}^2 + 2\mu\phi_{ij}\phi_{ij}) \quad \text{and} \qquad (6.72)$$

$$\mathscr{H}_D = -K \sum_{\mathbf{r} \neq \mathbf{r}'} \left[\mathbf{b}(\mathbf{r}) \cdot \mathbf{b}(\mathbf{r}') \ln\left(\frac{\mathbf{r} - \mathbf{r}'}{a}\right) - \frac{\{\mathbf{b}(\mathbf{r}) \cdot (\mathbf{r} - \mathbf{r}')\} \{\mathbf{b}(\mathbf{r}') \cdot (\mathbf{r} - \mathbf{r}')\}}{|\mathbf{r} - \mathbf{r}'|^2} \right]$$

$$+ E_c \sum_{\mathbf{r}} |\mathbf{b}(\mathbf{r})|^2 \; . \qquad (6.73)$$

Here $\mathbf{b}(\mathbf{r})$ is the dimensionless Burgers vector associated with the singularity. The summation in (6.73) is over a mesh of sites of spacing $\sim$ the dislocation core diameter, a. Moreover

$$K = \frac{\mu(\mu + \lambda)}{2\pi(2\mu + \lambda)} a_0^2 \; , \qquad E_c = (1 + C)K \; , \qquad (6.74)$$

where a_0 is the spacing of the underlying lattice, and C is a measure of the ratio a/a_0. The Burgers vectors satisfy the "charge neutrality" condition $\sum_{\mathbf{r}} \mathbf{b}(\mathbf{r}) = 0$. Using $\mathscr{H}_E$ to compute the partition function, one proceeds as outlined in Sect. 6.10.3.

The essential point is that the order parameter field must be modified to take explicit account of the strong perturbations due to the singular defects. It is important to recognize that the local deviations of the order parameter from the ideal value have two components, a smoothly varying one and a singular one. It is clear too, that this approach still falls short of qualifying as a theory based on first principles.

6.11 Overview

In this chapter we have seen how the Landau theory can be used to understand the occurrence of various structures. One must of course know (in advance) something about the structure concerned in order to make practical use of the theory. Such information is encoded in the order parameter, and essentially comes from experiment. This might give the erroneous impression that the Landau theory is not of much value. The various illustrations provided should be adequate to show that such an opinion would be unjustified. Especially where there is a clash of order parameters, the Landau theory provides a valuable perspective on the possible outcomes. And in cases like that of the blue fog in cholesterics (Sect. 6.9), it gives insight into how perfect ordering can be stalled by frustration effects.

At a more basic level, the Landau theory has been the fountainhead of important concepts such as symmetry breaking and the order parameter, in turn leading to the Goldstone modes (Sect. 3.4) and the homotopic classification of defects (Chap. 4).

The Landau theory has evolved continuously. One major development relates to the incorporation of the effects of fluctuations (Sect. 6.8). Extensions such as the de Gennes model (Sect. 6.6.3) have also been made involving two fields, namely $\psi(r)$, which relates to the distribution of the centres of gravity of the molecules, and $\delta n(r)$, which describes the orientations of the director. One may ask whether structures such as the DRP model and the CRN can be described in this manner, in terms of two fields related respectively to the *positions* of the basic building blocks (Fig. 2.11), and their *orientations*. Such a generalized theory for condensed matter does not exist at present, although there is one treatment [6.41] couched in the language of fibre bundles. There are also some gauge theories tailored for special cases [6.42, 43]. It should also be added that there is some doubt as to whether a basic principle like gauge invariance exists in condensed matter physics [6.44]. Nonetheless, the need for two types of fields for describing the tiling seems plausible.

In several phases of condensed matter it is not the order but the topological defects in that order, that is the dominant feature. Indeed, defects also control the phase transitions in many cases. In many treatments of defect-mediated transitions [6.36], the defect explicitly enters in the Hamiltonian. When they dominate the transition, we are essentially dealing with phase transitions of a pure gauge field [6.37].

It is worth noting that in the Landau theory the variables are coarse-grained. The theory lies between a full-fledged microscopic theory and a fully phenomenological macroscopic theory: it is a 'mesoscopic' theory. As such, certain details like the coordination number of individual atoms are not accessible. For such information, one must resort to the methods described in Chap. 5. Microscopic details form useful inputs for the Landau theory which, once formulated, can often lead to significant progress in our understanding of the condensed matter systems concerned.

7. Tilings in One Dimension

We now focus our attention on 1D tilings. While it is true that tilings in 1D are not as rich in structure as those in higher dimensions, the results to be obtained serve as valuable pointers to the possibilities in more involved situations.

The basic question is the following: Given identical atoms, in what different ways can they be arranged on a line? As stated above, the problem is ill posed, and additional conditions must be prescribed. Introducing suitable constraints, *Aubry* [7.1–3] and *Reichert* and *Schilling* [7.4, 5] have carried out some studies which are significant on account of their relation to dynamical maps [7.6–8]. Here we will see that results obtained in the study of dynamical systems can also be exploited in the study of structural problems [7.9].

7.1 Structures and Competing Periodic Potentials

7.1.1 The Problem

Consider a 1D harmonic chain of identical atoms in a periodic "substrate" potential (Fig. 7.1). The potential energy of such a system can be modelled as

$$\Phi(\{u_i\}) = \sum_i L_i \quad \text{where} \tag{7.1a}$$

$$L_i = \lambda V(u_i) + W(u_i - u_{i-1}) - \mu(u_i - u_{i-1}) \ . \tag{7.1b}$$

Here u_i is the position of the ith atom in the chain, $\lambda V(u_i)$, a periodic substrate potential of amplitude λ and periodicity $2a$, and $W(u)$, the harmonic, nearest-neighbour potential. The quantity μ is the line tension (pressure).

Formally, L_i can be regarded as a Lagrangian density and Φ interpreted as an action, with i representing discrete time. The analogy to mechanics goes even further, as we shall see. In passing we note that model (7.1) with

$$W(u) = \tfrac{1}{2}Cu^2 \ , \quad V(u) = \tfrac{1}{2}\left(1 - \cos\frac{\pi u}{a}\right) \tag{7.2}$$

is known as the *Frenkel–Kontorova* model, and is often used in the study of dislocations.

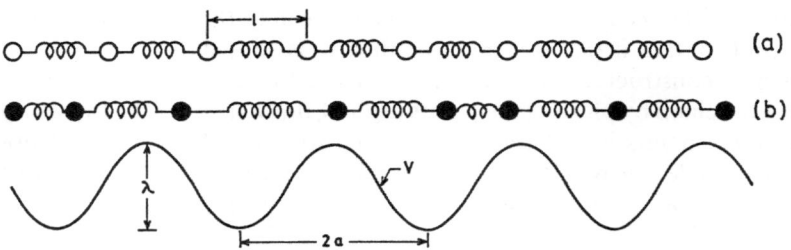

Fig. 7.1. (a) shows a periodic chain of identical atoms held by harmonic springs. When a substrate potential is present, the atom positions get disturbed as in (b), the disturbance being determined by the strength λ and the ratio $(l/2a)$

In the absence of the substrate potential V, the atoms in the chain are equally spaced (in equilibrium), the spacing l being determined by W and μ. But when V is switched on, each atom is under the influence of two competing periodic potentials. The structure which finally results depends on the one hand on the ratio $(l/2a)$ of the two periodicities, and on the other, on the amplitude λ of the substrate potential V. The various possible configurations of the chain are conveniently characterized by trajectories in a suitable phase space.

7.1.2 Structures and Maps

Our interest lies in the stationary configurations of (7.1) described by the equations

$$\frac{\partial \Phi}{\partial u_i} = \lambda V'(u_i) + W'(u_i - u_{i-1}) - W'(u_{i+1} - u_i) = 0 \ , \tag{7.3}$$

where the primes denote derivatives. Analytical solutions of (7.2) for $\lambda \neq 0$ are not easy to find, but several features of these solutions can be discovered and understood by employing techniques used in the study of dynamical systems [e.g. 7.6, 7]. In what follows, we will illustrate how such methods can be applied to the Frenkel–Kontorova model. Aubry has shown that these considerations can in fact be applied to the more general model of (7.1).

From (7.1–3) we obtain

$$u_{i+1} = 2u_i - u_{i-1} + (\lambda \pi/2Ca)\sin(\pi u_i/a) \ . \tag{7.4}$$

We may thus write

$$\begin{pmatrix} u_{i+1} \\ u_i \end{pmatrix} = \begin{pmatrix} 2u_i + (\lambda\pi/2Ca)\sin(\pi u_i/a) - u_{i-1} \\ u_i \end{pmatrix} \equiv T(\lambda)\begin{pmatrix} u_i \\ u_{i-1} \end{pmatrix} \tag{7.5}$$

where the $T(\lambda)$ is a nonlinear transformation mapping (u_i, u_{i-1}) to (u_{i+1}, u_i). In other words, given an initial pair of coordinates (u_1, u_0), the positions along the entire chain may be recursively generated via T, recall (2.18, 20) and (5.4). It is evident from (7.5) that if $\{u_i\}$ is a solution, then so is $\{u_i + 2a\}$.

Let us now construct a 2D "phase space" as in Fig. 7.2, and represent the structure of the chain by plotting the points $(u_0, u_1), (u_1, u_2), \ldots$. The trajectory formed by these points is a characteristic of the structure. However, it is not necessary to consider the whole of phase space to plot the trajectory. Rather, one may restrict attention to the square with vertices at $(0, 0)$, $(0, 2a)$, $(2a, 2a)$ and $(2a, 0)$, in view of

$$u_i = u_i \bmod 2a \ . \tag{7.6}$$

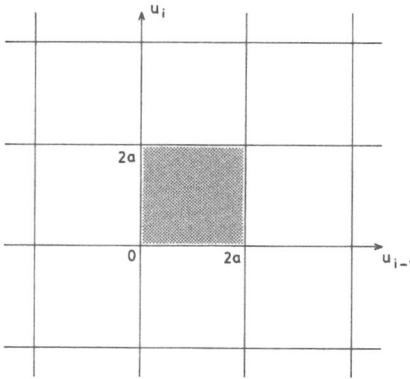

Fig. 7.2. Phase space for representing the structure of the 1D chain of Fig. 7.1. On account of the condition (7.6) it is sufficient to consider the shaded cell

Thus the transformation (7.5) can be regarded as the mapping of a square onto itself since $T(\lambda)$ takes some point within the square and moves it to some other position within the same square.

Digressing for a moment, we note that (7.5) is also studied by setting

$$p_i = u_i - u_{i-1} \ , \qquad \theta_i = u_i \bmod 2a \ , \tag{7.7}$$

whereupon (7.5) becomes

$$\begin{pmatrix} p_{i+1} \\ \theta_{i+1} \end{pmatrix} = \begin{pmatrix} p_i + (\lambda\pi/2Ca)\sin(\pi\theta_i/a) \\ p_{i+1} + \theta_i \end{pmatrix} \equiv \tilde{T}(\lambda) \begin{pmatrix} p_i \\ \theta_i \end{pmatrix} \ . \tag{7.8}$$

The transformation $\tilde{T}(\lambda)$ is referred to as the *standard map* which maps an infinitely long cylinder of circumference $2a$ onto itself.

7.1.3 Trajectories and Structures

Figure 7.3 shows some sample trajectories generated by $T(\lambda)$ corresponding to various values of the parameters of the potential, and various choices for the initial values (u_0, u_1). One sees a variety, ranging from an orbit which is a single point to an orbit which is a chaotic array of points. Let us consider the implications.

We start with the case where the trajectory $\{u_i\}$ is a single point P (say), i.e.,

$$T(\lambda)P = P \ . \tag{7.9}$$

This implies

$$u_{i+1} = u_i \, \text{modulo} \, 2a \ ,$$

i.e., the structure is crystalline, with periodicity $2a$.

Next consider a trajectory which is a sequence of s distinct points. If P is any point belonging to this set, then

$$T^s(\lambda) \, P = P \ , \tag{7.10}$$

i.e., the trajectory returns to P after every s steps. The chain is thus once again crystalline, this time with s atoms in the unit cell.

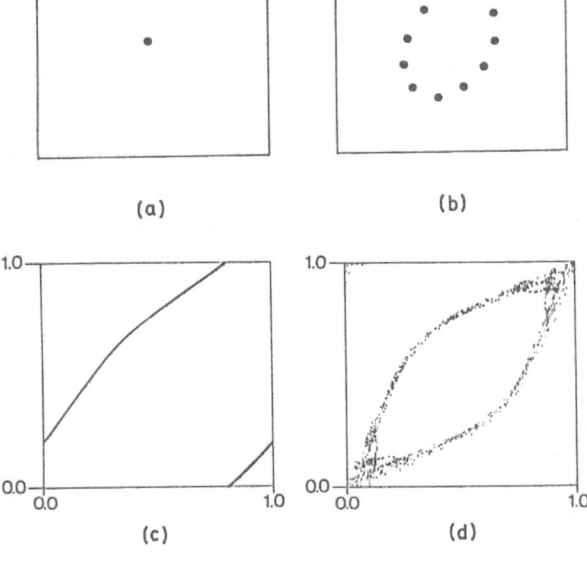

(a)

(b)

(c)

(d)

Fig. 7.3a–d. Trajectories generated by $T(\lambda)$ corresponding to representative values of C, λ, l and $2a$. (a) represents (7.9), and corresponds to a periodic structure; (b) the trajectory for a long-period structure and (c) that for an incommensurate structure. The orbit (d) is chaotic, and the corresponding real space configuration is random. However, such a configuration may not correspond to one of minimum energy

One can also have a trajectory like that in Fig. 7.3c which is dense on a smooth curve because the system never returns to a point already visited. The corresponding structure is incommensurate.

Other trajectories, which are chaotic, exist. They are the infinitely disconnected *Cantor sets* which are also dense. They correspond to amorphous structures, as we shall see in Sect. 7.3.

It is useful at this stage to draw parallels between the trajectories we have been considering and phase-space trajectories of dynamical systems. In dynamics one encounters three important kinds of orbits, corresponding respectively to periodic, quasiperiodic and chaotic motion. We observe there are equivalent orbits for the structural problem under study, with the evolution being in space rather than in time. In other words, the structure which emerges as i evolves (and one moves along) the chain can be periodic, quasiperiodic or chaotic.

7.2 Portrait of the Penrose Chain

In Fig. 7.3 we have seen an example of a trajectory associated with an incommensurate structure. It is natural to ask whether quasiperiodic chains (e.g., the 1D Penrose tilings) can also be accommodated in this framework. *Aubry* and *Godreche* [7.10] have proposed the following model of a 1D quasiperiodic chain. Consider a periodic chain of atoms located at sites $v_n = nl_0 + \alpha$ in a periodic potential with periodicity 2π such that $l_0/2\pi$ is an irrational number (Fig. 7.4). In order to create a system having the smallest possible energy, it is sufficient to remove those atoms which lie near the maxima of the potential within certain intervals of width Δ. The interval Δ is required to satisfy the condition $\Delta < l_0$ and $\Delta < 2\pi - l_0$ so as to prevent the occurrence of two (or more) consecutive vacancies. The occupied sites are now labelled x_n ($n = 0, 1, 2, \ldots$). Clearly, we must have $(x_n - x_{n-1}) = l_0$ or $2l_0$. The resulting sequence of tiles (or links) turns out to be quasiperiodic. This chain can be viewed as arising from the following *circle map*. Consider a circle of circumference 2π with an open region W (called a

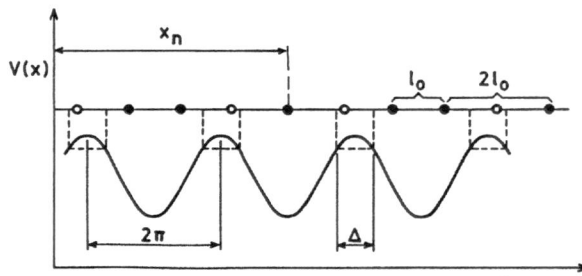

Fig. 7.4. The model of Aubry and Godreche. Black dots are atoms, open circles are vacancies

window) of arc length Δ. The map $x' = T(x)$ of this circle onto itself defined by

$$x' = \begin{cases} x + l_0 \, , & \text{if} \quad x + l_0 \notin W \\ x + 2l_0 \, , & \text{if} \quad x + l_0 \in W \end{cases} \tag{7.11}$$

generates this sequence $\{x_n\}$. *Aubry* and *Godreche* [7.10] assign a potential energy

$$\Phi(\{x_n\}) = \sum_n \cos x_n \tag{7.12}$$

to this structure and obtain the ground state of the model as an incommensurate periodic modulation of a given vacancy density. The concentration of vacancies (i.e., of tiles of length $2l_0$) is given by the ratio $\Delta : (2\pi - \Delta)$—a fact which follows from the uniform distribution of the successive images of a point on the circle. They have shown that the Penrose chain obtained by the projection method (Sect. 5.13) can also be generated by a circle map of the type (7.11). More precisely, the circumference of the circle is taken to be equal to $(1 + \tan \theta)$, which is the intercept of the y-axis with the strip $\mathcal{S}$ in Fig. 5.12, the window W is taken to have unit length, and l_0 is chosen to be $(1 + \tan \theta)^{-1}$. [Recall that $\cot \theta = \tau = (\sqrt{5} + 1)/2$.] Then the map is given by

$$x_n = \begin{cases} x_{n-1} + \cos \theta \, , & \text{if} \quad nl_0 \in W \\ x_{n-1} + \sin \theta \, , & \text{if} \quad nl_0 \notin W \, . \end{cases} \tag{7.13}$$

An alternative way of writing (7.13) is

$$x_n - x_{n-1} = [1 + (1/\tau)I_n] \sin \theta \, , \qquad \text{where} \tag{7.14}$$

$$I_n = \left\lfloor \frac{n}{\tau} \right\rfloor - \left\lfloor \frac{n-1}{\tau} \right\rfloor \, . \tag{7.15}$$

As is easily seen, I_n is either 0 or 1. Thus, as n increases, one stays in the same "sublattice" whenever $I_n = 0$ and changes sublattices whenever $I_n = 1$. We may re-scale x_i so as to absorb the factor $\sin \theta$ in (7.14).

Now any number x can always be written as

$$x = \lfloor x \rfloor + \{x\} \, , \tag{7.16}$$

where $\{x\}$ is the fractional part of x. This suggests a convenient phase space as in Fig. 7.5. The trajectory which results as n increases is also shown. There are clearly two disjointed segments, namely, ABCDA and EF. The two segments are associated respectively with the L and S sublattices. As n increases, the phase space point moves to new positions, remaining on the same branch if $I_n = 0$ and changing branches if $I_n = 1$. When $n \to \infty$, the trajectory becomes uniformly dense on the segments. Also there is a clear gap OE since there is a lower limit on

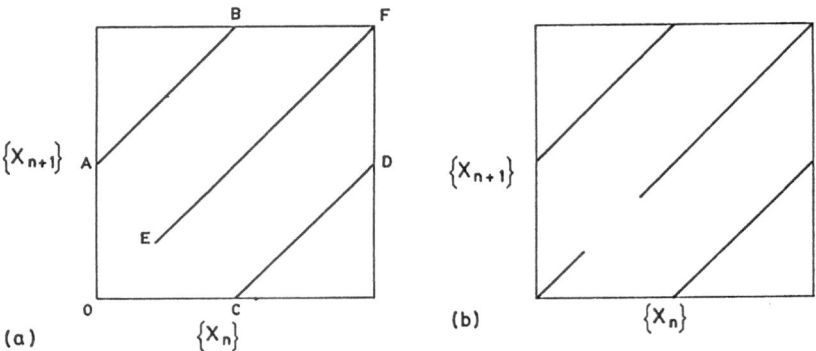

Fig. 7.5a, b. Trajectory plots for two Penrose chains belonging to the same local isomorphism class

$\{x_n\}$ for the S sublattice. This is a characteristic of the Penrose chain. A gap always occurs, whatever the choice of α and β in (2.15). However, the location of the gap may vary, as in Fig. 7.5b.

7.3 Spatial Chaos and Amorphous Structures

As already mentioned, *Reichert* and *Schilling* [7.4, 5] have examined whether spatial chaos implies amorphous structures. They approach the problem in two steps. To start with, they use purely mathematical arguments to construct a transformation T which generates a chaotic orbit. They then examine the pair correlation function $G(r)$ of the structure associated with such a chaotic orbit. Finding a strong resemblance between this $G(r)$ and that of an amorphous structure, they then construct a physical model leading to a chaotic orbit. Here we briefly review some highlights of these studies.

Consider a 1D chain with atoms at positions u_n such that

$$u_{n+1} - u_n \equiv v_n = A + Bx_n , \quad A > 0, \quad B > 0 . \tag{7.17}$$

Without loss of generality, we can set $u_0 = 0$. If the mean value of x_n is zero, then we have

$$\lim_{N \to \infty} \frac{1}{2N}(u_{n+N} - u_{n-N}) = A , \quad \text{for all } n . \tag{7.18}$$

Thus the mean particle spacing is A, and consequently the parameter B is a measure of the deviations of the nearest neighbour distances from their mean value.

A recursion rule for the generation of x is required now. For this purpose, Reichert and Schilling employ the Baker transformation (Fig. 7.6) which is defined as

$$T\begin{pmatrix} x \\ y \end{pmatrix} = \begin{cases} \begin{pmatrix} 2x \\ \frac{1}{2}y \end{pmatrix}, & 0 \leqslant x < \frac{1}{2} \\ \\ \begin{pmatrix} 2x - 1 \\ \frac{1}{2}(y + 1) \end{pmatrix}, & \frac{1}{2} \leqslant x < 1 \,, \end{cases} \tag{7.19}$$

where $0 \leqslant x < 1$, $0 \leqslant y < 1$. Starting with an initial point (α, β), x_n is the x-component of the nth iteration point of the Baker transformation reduced by $(1/2)$, i.e.,

$$x_n = (1, 0)\, T^n \begin{pmatrix} \alpha \\ \beta \end{pmatrix} - \frac{1}{2} \,. \tag{7.20}$$

It follows that v_n is restricted to the range $(A - B/2)$ to $(A + B/2)$. The ratio (B/A) is a crucial parameter, and is a measure of SRO. We assume $0 < (B/A) < 2$ to ensure $v_n > 0$. Figure 7.7 illustrates the evolution of the orbit and of the structure.

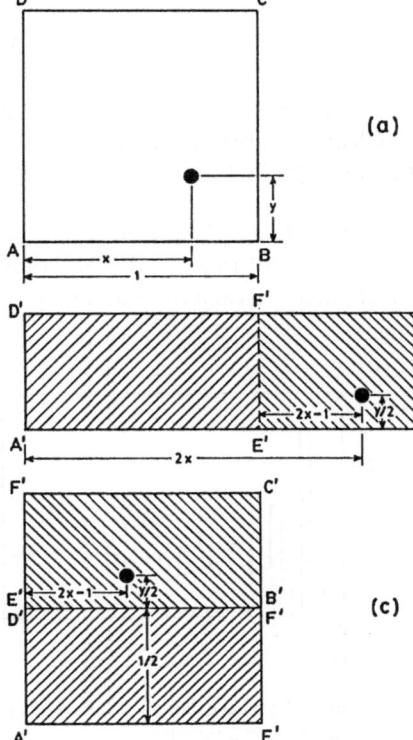

(a)

(b)

(c)

Fig. 7.6a–c. Illustration of the Baker transformation (7.19). The transformation is visualized in two steps. The unit square (**a**) is first deformed to be twice as large in length and half as large in height as in (**b**). The right half of the rectangle is then shifted to the top of the left half as in (**c**). The image of a representative point (*black dot*) is also shown

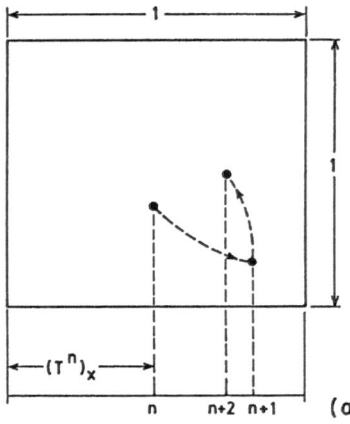

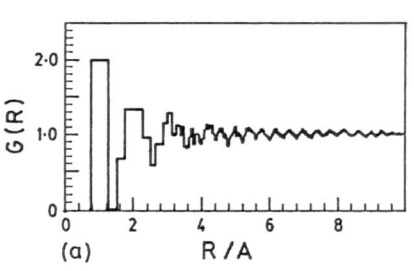

Fig. 7.7. (a) shows two stages in the Baker transformation. From these, x_n, x_{n+1} and x_{n+2} are obtained from (7.20) which then lead to v_n, v_{n+1} and v_{n+2} through (7.17). The resulting atomic configuration is sketched in (b)

To decide whether chaotic orbits as generated above do really represent amorphous structures, one must compute $G(r)$, the pair correlation function, which in turn is made up of the distribution functions for the various neighbours, i.e.,

$$G(r) = \sum_{j=1}^{\infty} G_j(r) \ . \tag{7.21}$$

Here G_j is zero outside the interval $[j(A - B/2), j(A + B/2)]$. Figure 7.8a shows a typical pair correlation function so obtained. Clearly, there is some resemblance to $G(r)$ for an amorphous solid. In particular, after the first few peaks, $G(r) \to 1$ as $r \to \infty$.

Encouraged by this result [7.4], *Reichert* and *Schilling* [7.5] have taken the next step of deducing such a behaviour from a physical model. The one they consider is an infinite chain of identical classical particles, each interacting up to

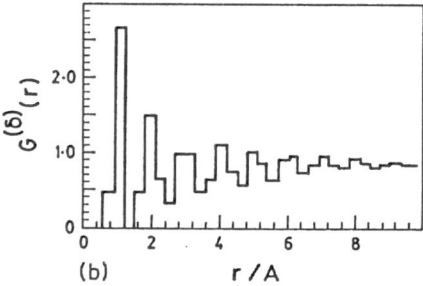

Fig. 7.8a, b. Amorphous-like pair correlation functions. (a) shows a typical example obtained using the Baker transformation, while (b) shows a similar one obtained using a potential model. [7.4, 5]

its rth nearest neighbour with potential energy

$$V(\{u_n\}) = \sum_n \sum_{l=1}^{r} V_l(u_{n+l} - u_n) , \qquad (7.22)$$

where V_l is the potential between the lth neighbours. Unlike the Frenkel–Kontorova model, (7.22) is translationally invariant in the sense that

$$V(\{u_n + a\}) = V(\{u_n\}) \quad \text{for all } a . \qquad (7.23)$$

The task now is to specify the V_l's, and look for stationary configurations. Here Reichert and Schilling note that at least one of the V_l's must be anharmonic for chaotic orbits to result. Also there must be frustration, which could be of geometrical, energetical or of any other origin. The simplest model for V_1 which achieves this is the potential

$$V_1(v) = \tfrac{1}{2}C_1 \{ [v - a_+ - a_- \sigma(v)]^2 - [c - a_+ - a_- \sigma(v)]^2 \} , \quad C_1 > 0 . \qquad (7.24)$$

As is evident, V_1 is made up of two parabolas patched together at c. Further $a_\pm = (a_2 \pm a_1)/2$, a_1 and a_2 being the positions of the minima in the two parabolas. The quantity

$$\sigma(v) = \text{sgn}(v - c) , \qquad (7.25)$$

and thus takes the values ± 1. For V_2, a harmonic potential is assumed:

$$V_2(v) = \tfrac{1}{2}C_2(v - b)^2 , \quad C_2 \gtreqless 0 . \qquad (7.26)$$

Figure 7.9 shows some representative situations for V_1 and V_2.

The equilibrium distances v_n for the model are given by the nonlinear difference equation

$$2\gamma v_n + v_{n-1} + v_{n+1} = \phi(v_n) , \quad \text{with} \qquad (7.27a)$$

$$\phi(v_n) = 2[b + (\gamma - 1)a_+] + 2(\gamma - 1)a_- \sigma(v_n) , \quad \text{and} \qquad (7.27b)$$

$$\gamma = 1 + C_1/2C_2 . \qquad (7.27c)$$

Observe that $\sigma_n = \sigma(v_n)$ acts like an Ising spin. Reichert and Schilling solve (7.27) making suitable assumptions about the sequence $\sigma = \{\sigma_n\}$ of the Ising variables. For instance, if $\sigma_n \equiv 1$, then a periodic system results. More interesting is the configuration $\{v_n\}$ obtained with a random sequences of $+1$'s and -1's, with probabilities p and $(1 - p)$ respectively. Such a sequence of σ's represents *quenched bond disorder*.

Figure 7.8b shows one of the pair distributions so obtained. Ignoring the many subtleties underlying the technique, it is enough for us to note that the pair distribution has similarities to that in Fig. 7.8a, confirming that spatially

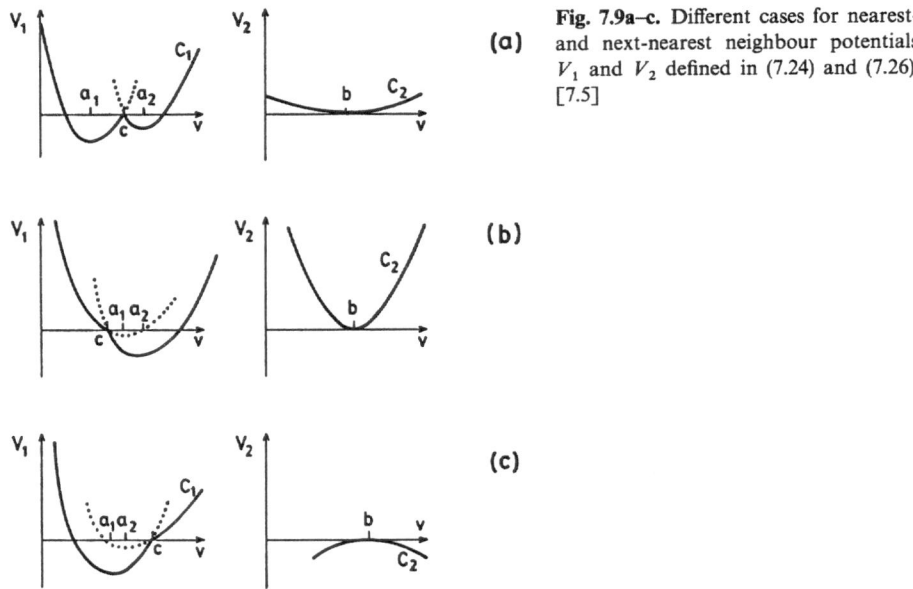

Fig. 7.9a–c. Different cases for nearest- and next-nearest neighbour potentials V_1 and V_2 defined in (7.24) and (7.26). [7.5]

chaotic structures can be derived from a (translationally invariant) potential model. We thus see that periodic, quasiperiodic, incommensurate and amorphous chains can all be deduced from a suitable potential.

7.4 Summary

In this chapter we have seen how packings in 1D can be studied by mapping techniques. Mathematically, the transformation T denotes an area preserving map; from a physical point of view it describes how the entire chain is built up, brick by brick. The effect of T is portrayed as a trajectory which is a signature of the structure generated. Depending on the values of the various parameters involved, different trajectories are possible. There are three basic types of trajectories, corresponding respectively to periodic, quasiperiodic and amorphous structures. The entire analysis is for structures at 0 K.

In crystallography one raises the question about the different periodic arrangements or patterns possible in a space of a given dimension. Here we have been discussing a complimentary question, namely, given building blocks of a particular type (e.g. atoms), what are the possible tilings, periodic and otherwise? Clearly the problem is of greater interest in the study of the noncrystalline state. It is an open question as to whether the methods reviewed in this chapter can be extended to structures in higher dimensions. Also, there is a need to consider the effect of nonzero temperature.

8. Ergodicity Breaking

We have seen earlier that many phases of condensed matter in thermal equilibrium are associated with the spontaneous breakdown of symmetry. In this chapter we deal with a more general concept called *broken ergodicity* [8.1]. It would seem that states of matter like glass, quasicrystals, etc. represent examples of ergodicity breaking.

8.1 Basic Ideas

Systems in thermodynamic equilibrium are ergodic. In the present context, ergodicity implies that the phase space trajectory which describes time evolution will come arbitrarily close to any specified point in the region of the phase space accessible to the system, given sufficient time. As a result, observed quantities are given by averages taken over *all of the allowed phase space*. In essence, ergodicity means that ensemble averages (i.e., phase space averages) can be replaced by time averages (over an infinite interval of time) of the variables evolving from any single initial condition.

Although one usually presupposes ergodic behaviour in statistical mechanics, many systems are in fact not ergodic *in practice*. In other words, their phase-space trajectories remain restricted to certain subsets of the allowed phase space for all *reasonable time scales of observation*. Therefore, if a time average is performed (allowing the system to evolve only during the time scale of observation), it would be equivalent to an ensemble average over only a limited region of phase space, and the result would differ from a full ensemble average. This absence of ergodicity is referred to as broken ergodicity.

8.2 Time Scales and Broken Ergodicity

Time scales play a crucial role, and merit further attention. In dealing with a physical system, we have to consider two basic types of time scales, one (τ_0) related to observations, and the other related to the slowest of the relevant dynamical processes in the system. One's assessment of the behaviour of the system depends very much on the relative magnitudes of τ_0 and the physical time scales.

Several scenarios are sketched in Fig. 8.1. In every case, the dynamical modes to the left of τ_0 (if any) may be regarded as "fast", while those to the right may be treated as "slow". Based on this operational scheme, one could, in favourable cases, define a "local" equilibrium state in which "all the fast things have happened and all the slow things have not".

A simple example will illustrate the point [8.1]. Suppose we add a little milk to a hot cup of coffee and observe the system continuously over long periods,of time. At first the milk will mix with the coffee and attain the temperature of the coffee; this will happen in a time ranging between a few seconds and a couple of minutes. Following this, over the next hour or so, the cup and its contents would slowly cool down to the temperature of the surroundings. Then, over a period of days, the entire contents of the cup would evaporate and mix with the air in the room. What is meant by equilibrium in this case? Clearly one can define various kinds of equilibrium, depending on which processes are declared fast and which are deemed to be slow.

The above discussion may be reinterpreted in terms of ergodicity, for which purpose we return to Fig. 8.1. In Fig. 8.1a, the time scale τ_0 is adequate for the phase space point to have visited a sufficiently representative sample of phase space so as to give the same averages as would be obtained for much longer times t of observation, including the limit $t \to \infty$. The system is therefore effectively ergodic, even though observations are made on a finite time scale τ_0. On the other hand, in Fig. 8.1c for example, there are many dynamical processes with time constants longer than τ_0, and in this case we have broken ergodicity. In general, we expect ergodicity to be broken if there are dynamical time scales much longer than τ_0. By convention, this term is used only when such time scales

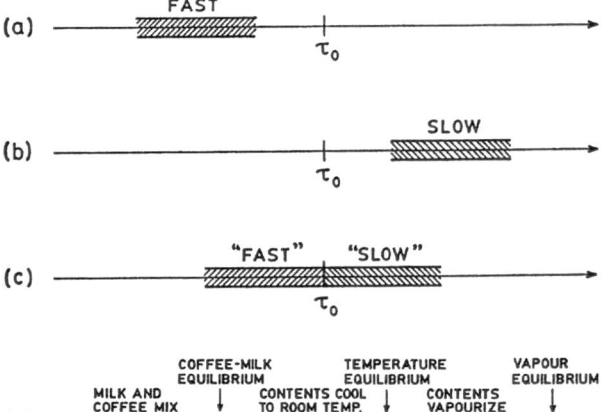

Fig. 8.1a–d. Role of τ_0 with reference to the description of system behaviour. The shaded patches indicate the time scales of the degrees of freedom present in the system. For explanations, see text

are very large, i.e., of the order of years and more. Thus, in the case of milk in coffee one would not speak of ergodicity breaking; but in the case of glass, one would.

8.3 Broken Ergodicity and Symmetry Breaking

When a system orders, it breaks symmetry. What happens, if anything, when ergodicity is broken, and is it in any way related to symmetry breaking? To sort this out, let us consider the 2D-Ising model.

Figure 8.2 shows schematically the phase space trajectories of this system in the disordered and the ordered states, as well as the relevant time scales. Above T_c, the system explores the whole of the allowable phase space Γ (more accurately, a sufficiently representative sample of it). Below T_c the system orders with net magnetization $\langle \Sigma S_i^z \rangle$ being either up or down, i.e., $\pm M$, say. The corresponding phase space trajectories are restricted to appropriate parts Γ^+ and Γ^- of Γ ($= \Gamma^+ \cup \Gamma^-$), over time scales less than a certain τ_0. There is the

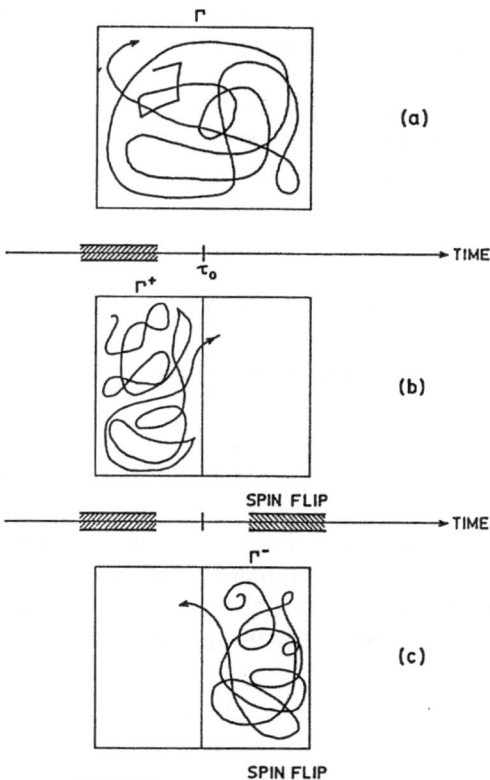

Fig. 8.2a–c. Schematic phase space trajectories for a spin system. (a) refers to the disordered state where the trajectory explores the whole of the phase space Γ. (b) corresponds to a spin up system, and on time scales $\tau < \tau_{\text{spin-flip}}$, the trajectory is confined to a region Γ^+ of Γ. Only for $\tau \gtrsim \tau_{\text{spin-flip}}$ does the trajectory leave Γ^+. (c) likewise refers to a spin down system

process of spin flip which can cause a macroscopic reversal of order from $\pm M$ to $\mp M$. Thus the confinement of the phase space trajectory to a single component Γ^+ or Γ^- applies only so long as $\tau_0 \ll \tau_{flip}$, the average time required for such a reversal. For $\tau_0 \ll \tau_{flip}$, there is broken ergodicity and local equilibrium (with respect to Γ^+ or Γ^-.) When $\tau_0 \gg \tau_{flip}$, the system explores the whole of Γ and ergodicity is restored. The point however, is that this would require τ_0 to be astronomically large when $T \ll T_c$.

The above example illustrates the relationship between broken symmetry and broken ergodicity. We see that ordering not only leads to symmetry breaking but also to ergodicity breaking. Nevertheless, the breaking of ergodicity in such a situation is not as significant as it might appear at first sight. Within a single component like Γ^+ or Γ^-, there is *internal ergodicity*, i.e., phase space averages equal time averages, and the usual apparatus of equilibrium statistical mechanics is effective when applied to one component at a time. Such an average $\langle \ \rangle_\alpha$, over a specific component Γ^α of Γ, is referred to as a restricted average [8.1, 2]. It implies equilibrium within the component concerned and not between components. This quasi-equilibrium must be contrasted with the conventional infinite-time equilibrium, which may be referred to as the Gibbs equilibrium. A quasi-average is quite adequate when dealing with broken symmetry. Thus, when calculating the specific heat of the ordered Ising magnet, it does not matter whether the ordering corresponds to $+M$ or $-M$; nor does spin flip make any significant contribution to this quantity.

Ergodicity breaking *without* symmetry breaking is possible, and does occur. In that case quasi-averaging alone would not do, as we shall discuss presently.

8.4 The Spin Glass

The spin glass [8.3] is an excellent example of a condensed matter system that lacks order (in the usual sense), but exhibits broken ergodicity. To introduce the idea of a (2D) spin glass, we first start from the ferromagnetic Ising model. The Hamiltonian of the latter is

$$\mathscr{H} = -J \sum_{\langle ij \rangle} \sigma_i \sigma_j \ , \tag{8.1}$$

where $J > 0$ is the nearest-neighbour interaction, and $\sigma_i = \pm 1$ is the Ising spin variable. The summation $\langle ij \rangle$ is over nearest neighbour pairs. Now suppose that, instead of being ferromagnetic, the exchange integral fluctuates *randomly* between the values $\pm J$, i.e., between ferro- and antiferromagnetic interactions. The Hamiltonian then becomes

$$\mathscr{H} = -J \sum_{\langle ij \rangle} \sigma_i A_{ij} \sigma_j \ , \tag{8.2}$$

where $A_{ij} = \pm 1$ is a random variable. In contrast to (8.1), there are now two variables: site variables $\{\sigma_i\}$ and bond variables $\{A_{ij}\}$. This system does not order at low temperatures, the various spins being *frozen* at random in up and down states owing to the competition between the ferro- and antiferromagnetic interactions. Such a disordered state with a frozen spin configuration is a convenient model for a spin glass. A particular configuration $\{A_{ij}\}$ of the bond variables will be denoted by A, and the totality of bond configurations will be denoted by $\{A\}$.

For a moment let us suppose that both the $\{\sigma_i\}$ and the $\{A_{ij}\}$ fluctuate rapidly on a time scale smaller than τ_0. One then refers to the system as an *annealed* spin glass, and the variables as *annealed variables*.

The partition function is given by

$$Z_{\text{ann.}} = \sum_{\{A\}} \sum_{\{\sigma\}} \exp(-\beta \mathscr{H}) \; , \tag{8.3}$$

where

$$-\beta \mathscr{H} = K \sum_{\langle ij \rangle} \sigma_i A_{ij} \sigma_j \; , \quad (\beta = 1/k_{\text{B}} T) \tag{8.4}$$

and $K = \beta J$. Observe that the sum is over both the spin as well as the bond fluctuations (Fig. 8.3).

True spin glasses correspond to the situation in which the $\{A_{ij}\}$ are not annealed random variables, but *frozen*, or *quenched* random variables, in contrast to the $\{\sigma_i\}$. The $\{A_{ij}\}$ do not undergo thermal fluctuations in the time scales of observational interest. Averages $\langle \; \rangle$ must therefore be evaluated in two steps, first by taking a thermal average as usual over the $\{\sigma_i\}$ corresponding to a given $\{A_{ij}\}$, followed by an average over the distribution of the $\{A_{ij}\}$ (Fig. 8.3). Symbolically,

$$\langle \; \rangle = \langle \langle \; \rangle_\sigma \rangle_A \; , \tag{8.5}$$

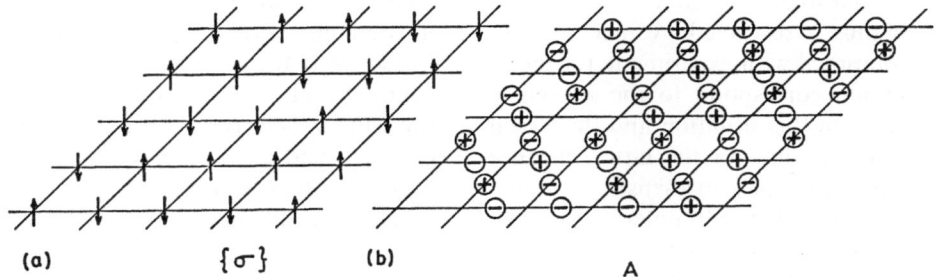

(a) $\{\sigma\}$ (b)   A

Fig. 8.3. Part (a) illustrates a particular spin configuration $\{\sigma\}$ with some spins up and others down. In (b) is shown a typical bond configuration A with some bonds A_{ij} being ferromagnetic $(+)$ and others antiferromagnetic $(-)$. While taking averages, one first sums over all spin configurations possible for a given A, and then averages over the distribution of the $\{A_{ij}\}$

where $\langle \ \rangle_\sigma$ is the *thermal* average over a given "component" Γ^A and $\langle \ \rangle_A$ is a *statistical* average over the distribution of the $\{\Gamma^A\}$.

Let us illustrate the above procedure by considering the case of the free energy. Taking a particular configuration A,

$$Z(A) = \sum_{\{\sigma_i\}} \exp\left(K \sum_{\langle ij \rangle} \sigma_i A_{ij} \sigma_j \right) . \tag{8.6}$$

The free energy $F(A)$ corresponding to the component Γ^A is

$$F(A) = -\beta^{-1} \ln Z(A) . \tag{8.7}$$

Let us now suppose that the bond probability $p(A_{ij})$ is given by

$$p(A_{ij}) = \begin{cases} \alpha , & \text{for} \quad A_{ij} = 1 , \\ 1 - \alpha , & \text{for} \quad A_{ij} = -1 . \end{cases} \tag{8.8}$$

The total probability $P(A)$ for the configuration A can then be expressed as

$$P(A) = \prod_{\text{all bonds}} p(A_{ij}) . \tag{8.9}$$

The free energy averaged over all possible bond configurations is therefore

$$F = \sum_{\{A\}} P(A) F(A) \Big/ \sum_{\{A\}} P(A) . \tag{8.10}$$

The two steps indicated in (8.5) are implemented in (8.6 and 10).

The usual ensemble averages we compute in statistical mechanics average over all thermal fluctuations. Owing to ergodicity, this is also equivalent to time averaging. In the spin glass problem, $\langle \ \rangle_\sigma$ is an average over the thermal fluctuations of the spin, but in a specific configuration A of the bond variables. It is thus a restricted average over Γ^A. Unfortunately, this average cannot be compared with experiment, for there is no reason why the experimental sample should correspond to the chosen bond configuration A. A subtle difference between this situation and the ferromagnetic Ising system considered earlier is worth noting. In the latter instance, though Γ^+ and Γ^- are different, they represent copies, in a sense. Averaging over a single component (either Γ^+ or Γ^-) does not give results different from those obtained by a full Gibbs average for any of the usual properties. In the case of the spin glass, however, the different bond configurations which can be realised are not (symmetry modified) replicas. A proper averaging over all the realizations of A, i.e., *configuration averaging*, is therefore necessary. Note, incidentally, that every frozen configuration A breaks ergodicity.

8.5 The Case of Glass

Glass is a classic example of broken ergodicity. As we know, there is no unique structure (i.e., configuration) for glass, and the structure is whatever the liquid finds itself trapped in. We attempt to show this schematically in Fig. 8.4 which is a plot of the potential energy V as a function in configuration space, i.e., the space of all coordinates collectively denoted by X [8.4]. In reality of course we would have to visualize V as a multidimensional surface with ridges, valleys, etc.

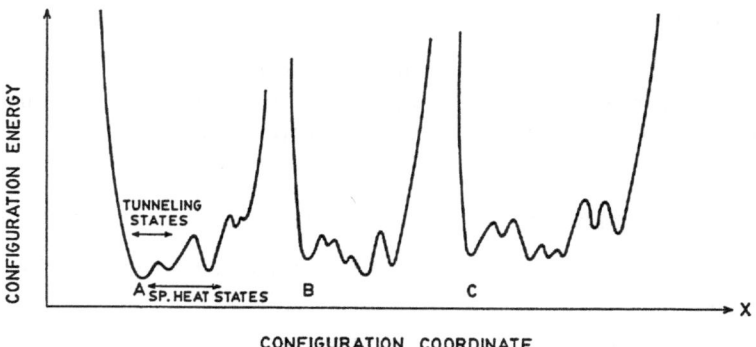

Fig. 8.4. Configuration energy as a "random mountain range". Explanations in text

When a given specimen of glass is formed by cooling, it finds itself trapped in one of the deep valleys such as A. From there, it has access, by tunneling as well as thermal fluctuations, to neighbouring low-lying states which do not have high barriers in between. However, access to states like B or C is ruled out. The glass A thus lives in its own world, Γ^A. Similarly, other specimens of the glass might represent a trapping in B, C, and so on, each with different tunneling and specific heat states accessible to them. It is this variety of realizable configurations which leads to configurational entropy. Moreover, a macroscopic specimen of glass has in it regions representative of a sufficient number of the different realizable configurations like A, B, C, . . . , and each of these regions is itself sufficiently macroscopic for thermodynamics to be applicable. One can thus appreciate the need for configuration averaging before the results of theory are compared with experiment. This "self-averaging" property of systems exhibiting quenched disorder is valid for most properties of physical interest. (This is so, as long as fluctuations play the role they conventionally do. Exceptions arise when fluctuation effects dominate, especially in low-dimensional systems).

8.6 Generalization

Frozen degrees of freedom occur in many systems, and the following is a formal scheme for performing averages in such cases [8.5].

Consider a system at a temperature T_i with no degrees of freedom frozen. However, it is conceivable that one can identify a set X of degrees of freedom, with a pronounced tendency to be slow or sluggish at lower temperatures, and another set Y which always remains rapidly varying. Let $\mathcal{H}(X, Y)$ denote the Hamiltonian. Then the probability density $P(X)$ of X at temperature T_i is

$$P(X) = \frac{\int \exp\{-\mathcal{H}(X, Y)/k_B T_i\}\, dY}{\int dX' dY' \exp\{-\mathcal{H}(X', Y')/k_B T_i\}} \ . \tag{8.11}$$

Now suppose the system is suddenly quenched to a much lower temperature T_f. The slow degrees of freedom will then be frozen at the values prevailing just before quench. The energy of the *quenched* system will now depend on these frozen values of X, i.e.,

$$\exp\{-F(X)/k_B T_f\} = \int dY \exp\{-\mathcal{H}(X, Y)/k_B T_f\} \ . \tag{8.12}$$

But, as far as experiments are concerned, what is pertinent is the configuration-averaged quantity

$$F_{\text{expt.}} = \int dX\, F(X)\, P(X) \ . \tag{8.13}$$

Of course, it is not always possible to explicitly identify the set of frozen variables; glass is an example! A system where the set X can be explicitly identified and where the analogues to the formulae of conventional statistical mechanics can be derived is referred to by *Edwards* [8.5] as the "Theorist's ideal glass".

Broken ergodicity is clearly an important concept which, in a sense, generalizes broken symmetry. However, many aspects of broken ergodicity have yet to be investigated more rigorously in the light of recent developments in the theory of dynamical systems.

9. Symmetry Breaking—A Second Look

Several years ago, *Pauling* and *Hayward* [9.1] presented a very interesting study of the architecture of molecules. We are here interested in the next stage, the architecture of the condensed state—i.e., the macroscopic structures that can be built out of large collections of atoms or molecules. The phases of condensed matter may be regarded as arising from an attempt to fill space homogeneously with given elementary entities, subject to certain constraints arising from basic physical principles. This leads to the wide variety of phases one observes in matter in the condensed state.

During the last century, mathematicians considered the various periodic arrangements possible in space. This led eventually to the discovery of space groups [9.2]. Such investigations have now gone beyond mere periodic, space-filling patterns to sphere packing in general, indeed in Euclidean spaces of arbitrary dimensions [9.3]. Our primary interest, on the other hand, has been in the tiling of E^3 alone, but with "tiles" of various shapes. Every packing arrangement breaks Euclidean symmetry in some manner or the other, and in Chap. 3 we discussed how different phases emerge when different symmetries are broken. We can now ask the question: What are the different possible ways in which Euclidean symmetry can be broken? It is evident that this is a much more general problem than the one with which crystallography is concerned.

9.1 Orbits and Strata in Crystal Physics

A crystal is invariant under a certain space group G. Therefore, in dealing with crystal physics, one should really use the irreducible representations (IR) of G, just as one uses the IR's of point groups in molecular physics. In practice, one exploits the fact that the focus is usually on a particular wavevector k, (e.g., in lattice dynamics or band-structure calculations). The group considered is a subgroup of G, relevant to the wavevector k. This group, denoted by $G(k)$, is referred to as the *group of the wavevector* in solid state physics; it is the *little group* of G corresponding to the IR labelled by k [9.4, 5].

Let us now consider Fig. 9.1, which shows the central Brillouin zone of a square lattice. Consider the wavevector k_1, with an associated little group $G(k_1)$. From k_1 we can derive other vectors k_2, k_3 and k_4 by group

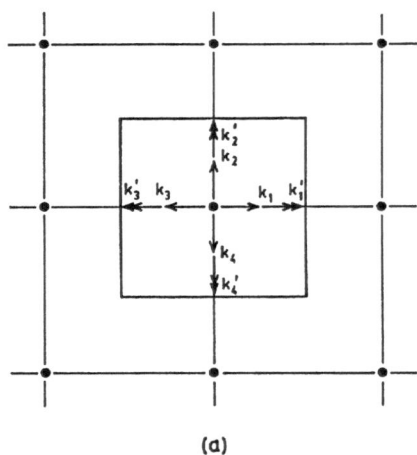

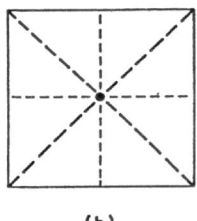

Fig. 9.1. (a) Central Brillouin zone of a square lattice, together with two orbits, namely, (k_1, k_2, k_3, k_4) and (k'_1, k'_2, k'_3, k'_4). (b) A few representative strata

(a)   **(b)**

action: (see Appendix C) e.g., $k_2 = C_4 k_1$, where C_4 is the rotation through $(2\pi/4)$. The set $\{k_1, k_2, k_3, k_4\}$ forms a *star* (Chap. 6); it is the *orbit* of k_1 under $G(k_1)$. The little groups $G(k_i)$ are conjugate with respect to each other; for instance,

$$G(k_2) = C_4 G(k_1) C_4^{-1} \; .$$

Consider the orbit $\{k'_1, k'_2, k'_3, k'_4\}$ in Fig. 9.1a. It is qualitatively similar to the one considered earlier; only the magnitude of the wavevectors is different. The union of like orbits is referred to as a *stratum*. Figure 9.1b illustrates a few strata.

Concepts like orbits, little groups and strata play an important role in symmetry breaking. Referring back to the example discussed in Sect. 3.1.3, the little group of the problem is nothing but the isotropy subgroup $G(\psi)$ associated with the state ψ. The set of states $\{\psi' | \psi' = g\psi, g \in U(1)\}$ is the family of ordered states, and constitutes an orbit. Each orbit is characterized by a particular value of $|\psi|$, the magnitude of the complex order parameter. The union of orbits with different values of $|\psi|$ constitutes a stratum.

9.2 Symmetry Breaking and Strata

We can now relate symmetry breaking to strata with the help of a simple illustration. Figure 9.2 is a family of free energy curves of the type shown in Fig. 3.2. This family is obtained from (3.1) by varying T across T_c. At every temperature, the system equilibriates at the minimum of the free energy. For $T > T_c$, there is a single valley corresponding to this minimum, but, at $T = T_c$, there is a bifurcation into two valleys.

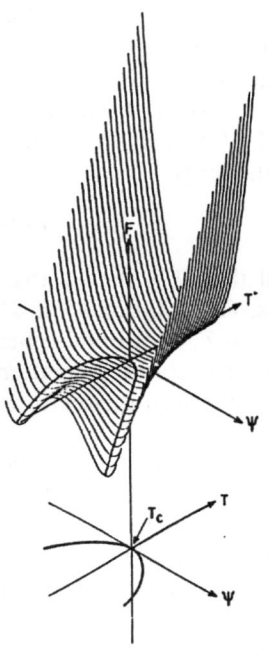

Fig. 9.2. A simple illustration of the strata in solution space. For $T > T_c$, the system remains in the only available stratum (union of all the $\psi = 0$ minima). At $T = T_c$ there is a change of stratum (which now is a union of all the pairs of minima). Correspondingly, there is a bifurcation and the system "flows" into (one of the branches of) the new stratum, leading to a phase transition

The set of minima at any given temperature T forms an orbit in the sense defined earlier. For $T > T_c$ the orbit consists of a single point, while for $T < T_c$, the orbit is a set of two points. Remembering that the union of all orbits of the same type constitutes a stratum, we see that there are two strata for the system described by the free energy functional (3.1).

The problem of symmetry breaking can now be stated as follows: Consider a G-invariant problem (e.g., the problem of finding the minima of a free energy functional which is invariant under a group G). Let $\mathscr{S}$ be the set of solutions of the problem (e.g., the set of equilibrium states). When one varies a parameter of the problem (e.g., the temperature), one follows a trajectory in the solution space $\mathscr{S}$. When this trajectory passes from one stratum to another, there is a change of symmetry [9.6].

9.3 Isotropy Subgroups of the Euclidean Group E(3)

We now return to the question posed earlier, and re-cast it as follows: Into which subgroups H can the symmetry G be broken, when G is the Euclidean group $E(3) = T(3) \wedge SO(3)$? (Here $\wedge$ denotes the semi-direct product.) This is a non-trivial mathematical question, requiring an analysis of the orbit structures of the various subgroups of E(3). In more technical language, one has to examine first whether the support of a measure on which E(3) acts is just one orbit, or a union of orbits, all of zero measure except one. The former of the two possibilities is the one which leads to the desired isotropy subgroups H [9.6]. It has been shown by *Kastler* et al. [9.7] that an isotropy subgroup H of E(3) exists only when $T_H (= H \cap T(3))$, i.e., the translational part of H, is of the form

$$\mathbb{R}^3, \quad \mathbb{R}^2 \otimes \mathbb{Z}, \quad \mathbb{R}^2, \quad \mathbb{R} \otimes \mathbb{Z}^2, \quad \text{or} \quad \mathbb{Z}^3 \,, \tag{9.1}$$

where $\otimes$ denotes the direct product of groups. (Recall that $\mathbb{R}$ is the additive group of real numbers, $\mathbb{Z}$ is the additive group of integers. The notation $\mathbb{R}^k$ or $\mathbb{Z}^k$ represents the direct product of k groups isomorphic to $\mathbb{R}$ or $\mathbb{Z}$. $\mathbb{R}^3$ is isomorphic to $T(3)$, $\mathbb{R}$ being isomorphic to $T(1)$). Are there physical states of matter characterized by isotropy subgroups H with T_H as in (9.1)? Indeed there are, as Table 9.1 shows [9.6].

Some explanatory remarks concerning this table are required. Let us take nematics as an example. As mentioned in Chap. 2, the centres of gravity of the molecules have only liquid-like order in this phase. However, the molecules are aligned on the average and the system has orientational order. The isotropy subgroup is $H = \mathbb{R}^3 \otimes D_{\infty h}$. Here $D_{\infty h}$ is the dihedral group consisting of all possible rotations about an axis, an inversion, and a horizontal mirror plane. Clearly $T_H = \mathbb{R}^3$, as shown in Table 9.1. H_0, the largest connected subgroup of H, is $\mathbb{R}^3 \wedge SO(2)$. We thus see that we can typecast H, H_0 and T_H purely from a group-theoretic angle, and it is pleasing that there are indeed realizations in Nature of these various possibilities for broken symmetry.

Informative as Table 9.1 is, it does not feature states such as incommensurate crystals, quasicrystals, etc. From Chap. 5 we know that incommensurate crystals and quasicrystals may be regarded as states with certain associated "internal coordinates". It would seem that, in these cases, it is not enough to consider (the

Table 9.1. Symmetry groups H of E(3) representing broken symmetry states. T_H denotes the translational part of H while H_0 is the largest connected subgroup of H. $\{e\}$ is the trivial group [9.6]

Family[a]	T_H	H_0	System
I	$\mathbb{R}^3$	$\mathbb{R}^3 \otimes U(1)$	Ordinary nematics
I	$\mathbb{R}^3$	$\mathbb{R}^3$	Exceptional nematics ($H = \mathbb{R}^3 \otimes D_{3h}$).
II	$\mathbb{R}^2 \otimes \mathbb{Z}$	$\mathbb{R}^3$	Cholesterics
II	$\mathbb{R}^2 \otimes \mathbb{Z}$	$\mathbb{R}^2 \otimes U(1)$	Smectics A
II	$\mathbb{R}^2 \otimes \mathbb{Z}$	$\mathbb{R}^2$	Smectics C
V	$\mathbb{R}^2$	$\mathbb{R}^2$	Chiral smectics C
III	$\mathbb{R} \otimes \mathbb{Z}^2$	$\mathbb{R}$	Rod lattices
IV	$\mathbb{Z}^3$	$\{e\}$	Crystals

[a] The family label follows the work of: M. Kleman, L. Michel: Phys. Rev. Lett. **40**, 1387 (1978).

breaking of) E(3) symmetry alone. Rather, one has to consider $E(3) \otimes \mathcal{G}_i$, where $\mathcal{G}_i$ is the symmetry group corresponding to the "internal" coordinates.

What are the various possibilities for $\mathcal{G}_i$? At present, there does not appear to be any systematic method of listing these. All one can do is to work backwards to deduce $\mathcal{G}_i$ whenever a new phase with internal degrees of freedom is discovered. There is, no doubt, some sort of analogy with the discovery of internal symmetries in particle physics.

Instead of dealing with composite groups $E(3) \otimes \mathcal{G}_i$, it seems that one could as well consider the isotropy subgroups of $E(n)$, $n > 3$. The projection scheme discussed in Chap. 5 is an example of this approach.

There are also states referred to as *ergodic*, concerning the definition of which there are different viewpoints [9.6–8], and these are yet to be classified. When they are, we would have an extension of Table 9.1.

9.4 More About Extensions to E(3)

So far we have tended to regard various phases of matter essentially as solutions to the tiling of E(3). That extensions of E(3) are required (even if we disregard possible "internal" symmetries $\mathcal{G}_i$) has been known for quite some time [e.g., 9.9]. In fact, even in the process of the enumeration of space groups, it was recognized that inversion symmetry must be considered explicitly. Thus G was taken to be

$$G = T(3) \wedge SO(3) \otimes \mathcal{I} , \qquad (9.2)$$

where $\mathcal{I}$ is the group of space inversion. In some space groups, the space-inversion symmetry is broken. With the advent of quantum mechanics, the group $SO(3)$ had to be enlarged to $SU(2)$, since wavefunctions are complex. This led to the discovery by *Kramers* [9.10] of a new invariance, namely, invariance under complex conjugation. Subsequently, *Wigner* [9.11] noted that the complex conjugation operation of Kramers was in fact the time-reversal operation T_R, which, at the wavefunction level, consisted of complex conjugation and the change $t \rightarrow -t$, simultaneously effected. Wigner's result brought one back to the classical form

$$G = T(3) \wedge SO(3) \otimes \mathcal{I} \otimes T_R , \qquad (9.3)$$

already familiar from Newton's equations. Finally, *Landau* [9.12] recognized that symmetry breaking based on (9.3) led to magnetic space groups.

Ordered magnetic structures are extensions of ordinary crystals. What about possible magnetic counterparts of other phases such as nematics, cholesterics, etc. enumerated in Table 9.1, or magnetic systems which are incommensurate? Indeed, examples of the latter are already known. So it seems that many more states remain to be classified. Only then can this aspect of condensed matter physics be said to be complete. Would such states exist in reality? The answer depends on whether there are (at least local) free energy minima corresponding to such states. Given the rich variety of structures already known, optimism about the existence of other states of unusual broken symmetry [based on (9.3)] is perhaps not unreasonable.

9.5 Patterns in Nonequilibrium Systems

The phases we have been discussing so far are those which occur in thermodynamic equilibrium. We recognize, however, that in systems far away from equilibrium there is a wealth of patterns, some of them strikingly similar to those of condensed matter in thermal equilibrium.

Figure 9.3 shows some patterns formed under nonequilibrium conditions [9.13–15]. All of them are associated with convection in a fluid held between

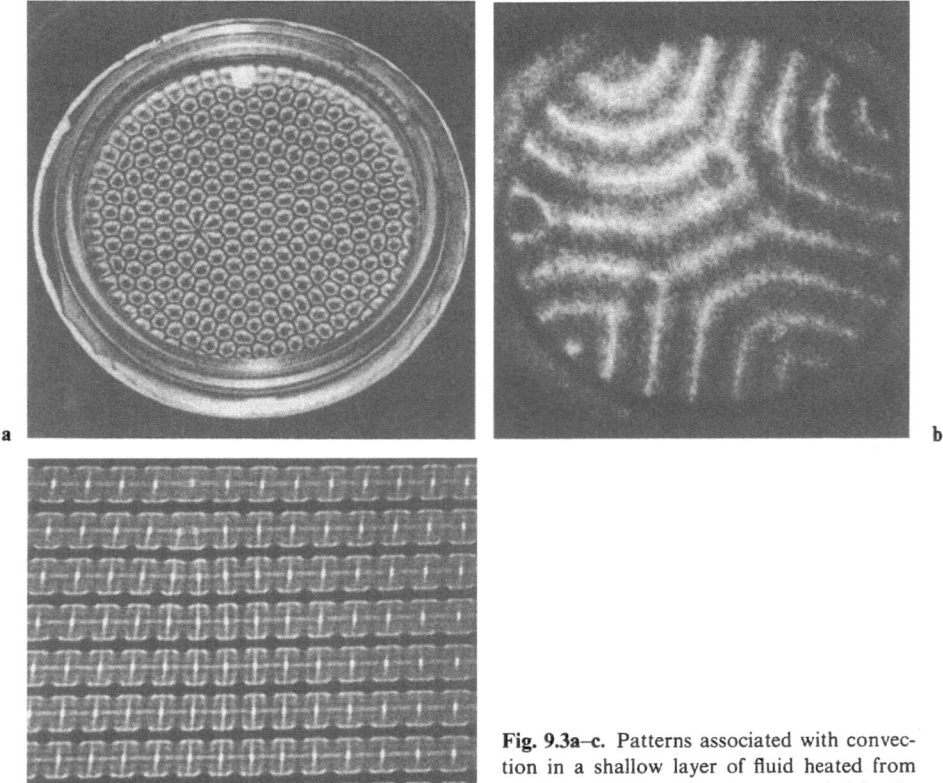

a

b

c

Fig. 9.3a–c. Patterns associated with convection in a shallow layer of fluid heated from below. Aluminium powder suspended in the fluid highlights the structure [9.13–15]

parallel plates and heated from below. One can clearly see not only regularity, but also well-marked defects in the patterns, as in equilibrium phases.

It would thus seem that symmetry breaking occurs also under nonequilibrium conditions. Indeed, *Sattinger* [9.16], among others, has discussed in detail the mathematical apparatus needed for analyzing such breakdowns of symmetry. Instead of the symmetry properties of the Hamiltonian, one has the symmetry group of an appropriate functional equation (e.g., differential equations in the case of fluid dynamics); instead of the symmetry group of a particular state of the Hamiltonian, one has the symmetry group of a particular solution obtained under specific boundary conditions. Barring such differences, there are strong parallels with the situation in equilibrium statistical mechanics. The problem is an entire subject in its own right, and is an area of active investigation at present. We do not pursue it further, as it is outside the scope of this book.

9.6 Cylindrical Crystallography

It is perhaps pertinent, at this juncture, to make a brief reference to an interesting simulation study performed by *Rivier* et al. [9.17] concerning 2D patterns with cylindrical or axial symmetry. One searches for space-filling structures with as much homogeneity as possible, the tiles or cells being as similar to each other and as isotropic as feasible, compatible with the boundary conditions. The aim is to describe the whole structure by a single algorithm which is as simple as possible.

One starts from a central core and proceeds outwards, placing points in the plane. These points are the centres of the cells concerned, and have (polar) coordinates

$$r(l) = a\sqrt{l} \ , \qquad \theta(l) = 2\pi\lambda l \ , \tag{9.4}$$

where $l = 1, 2, \ldots$ labels the individual cells, a is a typical linear dimension of a cell, and $\lambda(0 < \lambda < 1)$ is a parameter characterizing the structure. After the cell centres are located, the cells themselves are constructed by a Voronoi construction.

Figure 9.4 shows a few examples, illustrating how a simple parametrization like λ can encode widely different structures with cylindrical symmetry. Which one is selected in a particular context depends very much on what is being optimized (e.g., mechanical strength in the case of the spider's web, and the sharing of sunlight in the case of the daisy). Such selection can be studied only via a suitable variational principle.

The patterns of Fig. 9.4 are the outcome of an exercise in simulation. Can such patterns be deduced and classified by means of a broken-symmetry approach? The answer is not known at present. If indeed such a classification can be achieved, then one would truly have gone far in understanding the role of symmetry breaking vis-a-vis forms and shapes observed in Nature.

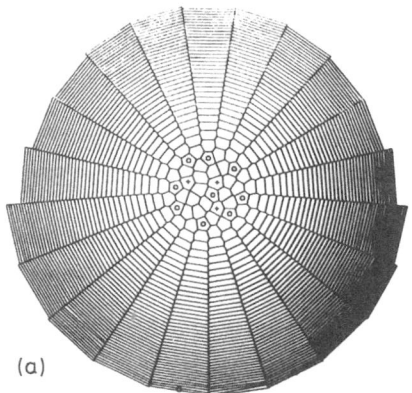

Fig. 9.4. The spider's web (**a**), the daisy (**b**), and the Bénard cell pattern (**c**), all generated with the same algorithm, (9.5), but with different values of λ

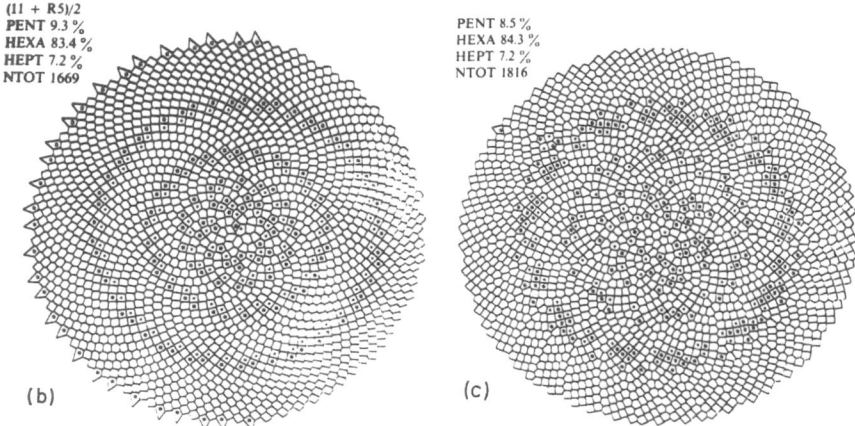

(11 + R5)/2
PENT 9.3 %
HEXA 83.4 %
HEPT 7.2 %
NTOT 1669

PENT 8.5 %
HEXA 84.3 %
HEPT 7.2 %
NTOT 1816

(b)   (c)

Finally, notwithstanding the slant in the foregoing chapters, we would like to reiterate that neither the phases of condensed matter in equilibrium nor the patterns it assumes under more general conditions is a matter of mere geometrical tiling. It is much more than that. As *Friedel* [9.18] has pointed out, " . . . the role of symmetry should not be over-emphasized at the expense of the physical content of the problems considered".

Appendix: Special Topics

It should be evident from the foregoing chapters that many new vistas are being opened up in the study of non-crystalline systems. While one cannot forecast as yet whether a comprehensive description of all such states of matter will eventually emerge, it is clear that several avenues must, in the meanwhile, be actively explored. And if past experience is any guide, tools of mathematics not totally familiar to condensed matter physicists, such as topology and differential geometry, will be increasingly needed in such explorations.

What follows is an assorted collection of addenda, somewhat mathematical in nature, intended to provide a minimum complement of techniques and concepts necessary to follow at least the *trends* in the literature (on the subject of present interest). Our selection of topics has clearly been influenced by the matter covered in the main text. No claim is made as to the set of addenda being complete in any sense. The main objective is to facilitate further study and enquiry by the motivated reader.

A. Hydrodynamics

Macroscopic properties of condensed matter are conveniently described within a formalism known as hydrodynamics. When a many-body system with a large number of degrees of freedom is disturbed from thermal equilibrium, almost all degrees of freedom relax rapidly to their equilibrium values. The few remaining ones—the so-called collective modes or hydrodynamic modes—relax slowly compared to the characteristic microscopic collision time of the system. The characteristic time of a typical hydrodynamic mode varies as some power of its wavelength. Such modes arise in the system either due to conservation laws and/or, due to *continuous* broken symmetries [A.1]. The occurrence of sound waves in a liquid is due to (local) conservation of energy, momentum and density. The frequency of this wave is linear in the wave number. The long-wavelength spin wave in an isotropic antiferromagnet is a manifestation of the breakdown of the continuous rotational symmetry of the spin system. It is essential that the broken symmetry be a continuous one, since this ensures that the system has the same energy for all orientations of the staggered spin arrangement relative to the lattice. Liquid crystals, superfluids and crystals are some other examples of ordered systems in which hydrodynamic modes arise owing to the existence of continuous broken symmetries.

It is the aim of hydrodynamics to give a description of a macroscopic system consisting of a large number of degrees of freedom in terms of a small set $\{\chi_a | a = 1, \ldots N\}$ of relevant (hydrodynamic) variables in the long-wavelength, low-frequency regime. If τ denotes a characteristic time of microscopic interactions and l denotes a characteristic microscopic length of the system, then the hydrodynamic regime is confined to frequencies ω and wavenumbers k satisfying the conditions $\omega\tau \ll 1$ and $kl \ll 1$. To appreciate these conditions, consider a cholesteric liquid crystal where the pitch of the helical arrangement of rod-like molecules is of the order of 5000 Å. Thus in this system, $l \sim 5000$ Å and the wavelength of a hydrodynamic excitation must be much larger than 5000 Å. A light scattering experiment probing the hydrodynamic regime will need infrared light. In contrast, a smectic liquid crystal has layers of rod-like molecules spaced at a typical distance ~ 10 Å, and hence visible light is quite adequate to probe the system. The quantities of experimental interest are correlation functions involving the $\{\chi_a\}$ or their Fourier transforms in (k, ω) space. The correlation function of interest in a light scattering experiment is the density–density correlation.

The domain of application of hydrodynamic theories is really vast. Systems in equilibrium, phase transitions involving equilibrium systems, systems far from equilibrium passing through instabilities like the Rayleigh–Benard instability in a fluid [A.2] and exhibiting the so-called dissipative structures [A.3], systems exhibiting chaotic behaviour [A.2], are all amenable to studies making use of the hydrodynamic formalism. The area of systems far from equilibrium, an active field of research, calls for the inclusion of the right type of nonlinearities in the hydrodynamic equations. The nonlinearity may be of reversible or irreversible origin. The reversible nonlinearities are those which do not contribute to the production of (local) entropy in the system and are relatively easier to incorporate. The irreversible nonlinearities are somewhat more difficult to incorporate into the hydrodynamic equations.

The purpose of this chapter is to describe how to set up the hydrodynamic equations of a given system. As the hydrodynamics of any ordered medium is merely a generalization of that of a simple fluid, we start with a simple fluid first. The scheme of generalization to include systems with broken symmetries is also dealt with. The example of a crystal, which has three continuous broken symmetries corresponding to translations, is worked out. In addition to the classical method of deriving hydrodynamic equations [A.4, 5], the more efficient method of Poisson brackets [A.6] is also described. Following this, the procedure of extending these hydrodynamic equations when topological defects such as dislocations are present, is described.

In equilibrium (and away from phase transitions), the linearized hydrodynamic equations suffice for the calculation of the correlation functions [A.7]. This is what the Landau–Lifshitz theory of fluctuating hydrodynamics [A.8] provides. There are also other approaches based on Fokker–Planck equations and master equations [A.9]. Away from equilibrium, as in the case of a fluid in large temperature gradients (Bénard convection regime) or near critical points, the calculation of correlation functions must be based on nonlinear equations [A.10]. In this Appendix, we shall not dwell upon such intricacies.

A.1 Hydrodynamic Equations

Since the hydrodynamic description of all ordered media is a generalization of the hydrodynamics of a simple fluid, we consider first a single component isotropic fluid.

The starting point is the identification of hydrodynamic variables $\{\chi_a(x, t)\}$. In a fluid, these are five in number: The mass density $\varrho(x, t)$, the energy density $\varepsilon(x, t)$ (or entropy density $s(x, t)$) and the three Cartesian components $g_k(x, t)$ $[= \varrho(x, t) v_k(x, t)$, v being the velocity] of the momentum density. One supposes that these variables form a complete set and that all the properties of the fluid can be expressed in terms of these variables—an assumption which cannot be proved. Here x is regarded as a "point" occupied by a material "particle". In a continuum description, the point (or the particle) consists not of just one atom of the fluid, but a sufficiently large number of atoms confined within a volume element ΔV (surrounding the point x) whose linear dimensions are much larger than typical interatomic distances, but small enough to the size of the system. Thermodynamic quantities such as pressure, entropy, etc. are well defined within this volume element. Thus the energy density $\varepsilon(x, t)$ at the cell whose centre of gravity is at $x \in \Delta V$, is practically constant within the region ΔV, but changes from one cell to another continuously. The density variables $\{\chi_a(x, t)\}$ are coarse-grained slow variables, i.e., one can imagine a relation

$$\chi_a(x, t) = \frac{1}{\Delta V} \int_{\Delta V} d^3x' [\chi_a(x', t)]_{\text{micro}}$$

where $[\chi_a]_{\text{micro}}$ are exact microscopic densities, e.g.

$$[\varrho(x, t)]_{\text{micro}} = \sum_l m \, \delta(x - x(l)) \; ,$$

and $x(l)$ is the position of the lth atom (of mass m). Alternatively (perhaps this is more satisfactory), one can regard $\chi_a(x, t)$ as an average of $[\chi_a(x, t)]_{\text{micro}}$ with respect to a *local* statistical mechanical distribution function [A.11].

The hydrodynamic description of a system is firmly rooted in thermodynamics. The main assumption, known as the *local thermodynamic equilibrium* (LTE), is that the laws of thermodynamics remain valid locally in the region ΔV. Thus the local entropy density $s(x, t)$ is the same function of the local thermodynamic variables as the equilibrium entropy is of the equilibrium thermodynamic variables.

The steps leading to the setting up of the hydrodynamic equations are the following. Details are given in the next section.

1) Choose the right variables $\{\chi_a\}$. The basic principles guiding the choice of such variables are conservation laws and the existence of continuous symmetry breaking.

2) Set up the equations of continuity (also known as the balance equations) expressing the local conservation of the quantities $\{\chi_a\}$. When χ_a are densities corresponding to extensive quantities, these equations are of the form

$$\frac{\partial}{\partial t} \chi_a + \nabla \cdot J_a = \sigma_a \ . \tag{A.1}$$

Here J_a, which is a vector or a tensor of appropriate rank, is the flux associated with χ_a; it has a reversible (J_a^R) and a dissipative (J_a^D) part, i.e.,

$$J_a = J_a^R + J_a^D \ .$$

σ_a is the source density associated with χ_a. The term $\nabla \cdot J_a$ denotes the appropriate divergence of J_a and depends on the tensorial nature of χ_a. When χ_a is the energy density, the LTE hypothesis is used to write the first law of thermodynamics expressing the conservation of energy in the form (A.1).

3) Write down the Gibbs relation of thermodynamics in the local equilibrium form (implementing the LTE hypothesis). This, together with the equation of continuity of entropy density, helps to determine the functional form of J_a^D. The fact that σ_a does not contain contributions from J_a^R determines the structure of J_a^R (the constitutive relation) to some extent. This is especially useful for the nonlinear part of J_a^R.

4) Invoke the continuity equation for the entropy density $s(x, t)$:

$$\frac{\partial}{\partial t} (\varrho s) + \nabla \cdot \left(\frac{q}{T} + \varrho s v \right) = \sigma_s \ , \tag{A.2}$$

where q is the heat flux vector and T is the local temperature. The entropy production density σ_s is of the form

$$\sigma_s = \sum_a{}' J_a^D \cdot X_a + q_k \frac{\partial}{\partial x_k} \left(\frac{1}{T} \right) , \tag{A.3}$$

where X_a is the generalized thermodynamic force associated with the variable χ_a, and the prime over summation means that the contribution due to the variable ε is excluded. The product $J_a^D \cdot X_a$ denotes an appropriate contraction of tensors.

5) Set up the Onsager relations [A.4, 5]

$$J_a^D = L_{ab} X_b \tag{A.4}$$

with the appropriate tensors L_{ab} of transport (or kinetic) coefficients. The summation over the index b implies that the flux of χ_a could couple to all the appropriate forces.

6) Now use the (thermodynamic) equation of state to express any additional thermodynamic variable (like pressure) in terms of the variables already chosen as hydrodynamic.

7) Substitute the expressions for J_a^R, J_a^D and the equations of state in (A.1). This completes the setting up of the hydrodynamical equations.

We now describe the detailed procedure. The motion of the fluid is completely specified by the trajectories of all its particles

$$x = x(a, t) ,$$

where the coordinates a also label a particular particle at the initial time $t = 0$, i.e. $a = x(a, 0)$. In classical hydrodynamics one conventionally uses (a, t) as independent variables. This is the *Lagrangian description* which might also be referred to as the rest-frame description since the coordinate frame is attached to the fluid particle. a. There is another description using the variables (x, t)—the so-called *Eulerian description* which is nothing but a description of dynamics in the laboratory frame. Let $\xi(x, t)$ be any fluid property (such as its density $\varrho(x, t)$; it could even be a component of a vector or a tensor property) which has a different functional form $\tilde{\xi}(a, t)$ when expressed in the rest frame. The time derivatives in the two frames are related as follows:

$$\dot{\xi}(x, t) = \frac{\partial \tilde{\xi}}{\partial t}\bigg|_{a = \text{constant}} = \frac{\partial \xi}{\partial t}\bigg|_{x = \text{constant}} + \frac{\partial x_k}{\partial t}\frac{\partial \xi}{\partial x_k} ,$$

or,

$$\dot{\xi} = \frac{\partial \xi}{\partial t} + v_k \frac{\partial \xi}{\partial x_k} \equiv \partial_t \xi + v_k \partial_k \xi$$

$$\equiv \xi_{,t} + v_k \xi_{,k} .$$

The notation commonly used to denote the derivative in the rest frame is

$$\frac{D\xi}{Dt} = \frac{\partial \tilde{\xi}}{\partial t}\bigg|_{a = \text{constant}} \tag{A.5}$$

Then we have the relation (abbreviating D/Dt as D_t)

$$D_t \xi = (\partial_t + v_k \partial_k)\xi . \tag{A.6}$$

Here D_t is called the *hydrodynamic derivative* or the *material derivative*. Two important properties[1] of this derivative are:

$$(1) \quad D_t(\xi\eta) = (D_t\xi)\eta + \xi(D_t\eta) , \tag{A.7}$$

$$(2) \quad D_t \int_V \xi(x, t) d^3x = \int_V (D_t\xi + \xi\partial_k v_k) d^3x . \tag{A.8}$$

Here ξ and η are any two continuous functions of (x, t).

With this background, the hydrodynamic equations governing the time evolution of the variables $(\varrho, v_k, \varepsilon)$ are set up in the laboratory frame using the principles of conservation of mass, momentum and energy. The law of mass conservation can be stated as

$$D_t m(t) = 0 , \tag{A.9}$$

where $m(t) = \int_V \varrho(x, t) d^3x$ is the mass-content of a region of volume V at an instant t. Then, using (A.5), we have

$$D_t \int_V \varrho \, d^3x = \int_V (D_t\varrho + \varrho \, \partial_k v_k) d^3x = 0$$

implying (since V is arbitrary) the local form of mass conservation:

$$D_t\varrho + \varrho \, \partial_k v_k = 0 . \tag{A.10}$$

This relation can be written alternatively as

$$\partial_t\varrho + \partial_k g_k = \partial_t\varrho + \boldsymbol{V} \cdot \boldsymbol{g} = 0 \tag{A.10'}$$

by using (A.6). Comparing this with the form (A.1), the current J_ϱ is seen to be equal to g; σ_ϱ is zero since there is no mass production or destruction.

To derive the expression for momentum conservation (when no external force is acting on the fluid), consider a closed surface Σ, described by the unit (outward) normal $\hat{n}$ in the rest frame enclosing a volume V. Let $\boldsymbol{T}(\hat{n}, x, t)$ be the (internal) force distribution per unit area on Σ, i.e., $\boldsymbol{T}$ describes the action of the fluid outside Σ on the fluid inside Σ. (The interior of Σ will be referred to as the region Σ.) In terms of the stress tensor σ_{ik}, the force is

$$T_i = -\sigma_{ik} n_k . \tag{A.11}$$

Newton's law relates the rate of change of the net momentum of the fluid enclosed in the region Σ to the net surface force $-\oiint_\Sigma \sigma_{ik} n_k \, d\Sigma$, where $d\Sigma$ is an area element of Σ:

$$D_t \int_V \varrho(x, t) v_i(x, t) d^3x = -\oiint_\Sigma \sigma_{ik} n_k \, d\Sigma . \tag{A.12}$$

---

[1] Equation (A.5) is trivial whereas (A.6) makes use of the fact that the volume element d^3x is related to the element d^3a by a time-dependent Jacobian $|\partial x_i/\partial a_k|$. For details, refer to *Malvern* [A.12].

Now, using (A.8) in the lhs and the Gauss divergence theorem in the rhs, and then equating the integrands, we obtain

$$D_t g_i + g_i \partial_k v_k + \partial_k \sigma_{ik} = 0 \ . \tag{A.13}$$

An alternative form of the above relation is (using A.6)

$$\partial_t (\varrho v_i) + \partial_k \Pi_{ik} = 0 \ , \tag{A.13'}$$

where $\Pi_{ik} = \varrho v_i v_k + \sigma_{ik}$ is known as the *momentum flux tensor*. Once again there is no source term since we suppose that no external forces are acting on the fluid.

In order to derive the local form of energy conservation we must assume that the laws of thermodynamics remain valid within a small volume element moving along with the fluid, i.e., we use the LTE hypothesis. The local form of the first law of thermodynamics is

$$D_t E = D_t U + D_t K = D_t Q + D_t W \ , \tag{A.14}$$

where E, U, K, Q and W are respectively the total energy, the internal energy, the kinetic energy, the heat input and the work done on the given region, and all quantities refer to the region of volume V contained within the closed surface Σ introduced earlier. The expressions for U and K are

$$U = \int_V \varrho u \, d^3x \ , \tag{A.15}$$

$$K = \tfrac{1}{2} \int_V \varrho v_k v_k \, d^3x \ , \tag{A.16}$$

where $u(x, t)$ is the internal energy per unit mass, and

$$E = \int_V \varepsilon \, d^3x \ . \tag{A.17}$$

The heat input $D_t Q$ is related to the heat flux q:

$$D_t Q = - \oiint_\Sigma q \cdot \hat{n} \, d\Sigma \ , \tag{A.18}$$

and the power input $D_t W$ is related to the surface traction T:

$$D_t W = - \oiint_\Sigma \sigma_{ik} n_k v_i \, d\Sigma \ ; \tag{A.19}$$

both of these transform into the following forms when use is made of the Gauss divergence theorem:

$$D_t Q = - \int_V q_{k,k} \, d^3x \ . \tag{A.20}$$

$$D_t W = - \int_V (v_i \sigma_{ik})_{,k} \, d^3x \ . \tag{A.21}$$

Substituting (A.17, 20, 21) in (A.14) and using (A.8), we obtain the two alternative forms of the conservation equation for ε:

$$D_t\varepsilon + \varepsilon v_{k,k} + (v_i\sigma_{ik} + q_k)_{,k} = 0 \tag{A.22}$$

$$\partial_t\varepsilon + (\varepsilon v_k + v_i\sigma_{ik} + q_k)_{,k} = 0 \ . \tag{A.22'}$$

So far, in the derivation of the conservation equations (A.10, 13, 22) we have not made any particular choice of the stress tensor σ_{ik}. Hence these equations hold for any continuous medium. The tensor σ_{ik} may in certain cases have an antisymmetric part. However, it is known [A.1, 13, 14] that if local torques are not allowed in the system, (i.e., if angular momentum is conserved) then the most general form of σ_{ik} must be

$$\sigma_{ik} = \tfrac{1}{2}(\sigma_{ik} + \sigma_{ki}) + f_{ikl,l}$$
$$f_{ikl} = -f_{kil} \ ,$$

i.e., the antisymmetric part of σ_{ik} can have at most a divergence term.

Equations (A.10, 13, 22) do not suffice to solve for the basic variables $(\varrho, v_k, \varepsilon)$ since the functional forms J_a (i.e., $\underline{\sigma}$ and $\boldsymbol{q}$) are not yet known. For this purpose, one invokes the Gibbs relation of thermodynamics, which brings along with it an additional variable—the entropy. Like the variables $(\varrho, v, \varepsilon)$, the entropy density s also satisfies an equation of continuity and has its associated flux $\boldsymbol{J}_s$ and a source density σ_s. As we shall show shortly, the substitution of conservation equations (A.10, 13, 22) for $(\varrho, v, \varepsilon)$ in the Gibbs relation determines $\boldsymbol{J}_s$ and σ_s. According to thermodynamics, σ_s is given by the expression (A.3) and J_a^D is linearly related to the forces X_a via the Onsager relation (A.4) (this relation is actually valid only for "small" departures from equilibrium). This fact, together with symmetry considerations, determines the functional forms of $\{J_a^D\}$. The fact that $\{J_a^R\}$ do not contribute at all to σ_s determines the functional forms of $\{J_a^R\}$ to some extent.

We now start with the conventional form of the Gibbs relation and then implement the LTE hypothesis to transform it into its local equilibrium form. Let S be the entropy contained in volume V inside Σ and let $M = \varrho V$ be the mass content of Σ. According to thermodynamics, $U = U(S, V, M)$, and we have the Gibb's relation (neglecting shear deformation of the continuum)

$$dU = \left(\frac{\partial U}{\partial S}\right)_{V,M} dS + \left(\frac{\partial U}{\partial V}\right)_{S,M} dV + \left(\frac{\partial U}{\partial M}\right)_{S,V} dM$$
$$\equiv T\,dS - P\,dV + \mu\,dM \ , \tag{A.23}$$

where T, P and μ are the temperature, pressure and chemical potential respectively. The fact that U is a homogeneous function of its arguments is expressed by the Euler relation

$$U = TS - PV + \mu M \ . \tag{A.24}$$

This relation is used to eliminate a variable, say the pressure P. In terms of the variables ϱ, $\varrho s = S/V$ and $\varrho u = U/V$, we rewrite relations (A.23, 24) as

$$d(\varrho u) = T d(\varrho s) + \mu d\varrho \ . \tag{A.25}$$

$$\varrho u = \varrho s T - P + \mu \varrho \ . \tag{A.26}$$

An alternative form of (A.25) is

$$du = T ds + (P/\varrho^2) d\varrho \ . \tag{A.27}$$

It is also possible to transcribe the above relations using E (or ε) instead of U (or ϱu). Noting that $E = K + U$ and that K is a homogeneous function of M and V, and using (A.15–17), we obtain the following relations:

$$d\varepsilon = T d(\varrho s) + \mu d\varrho + v_k \, dg_k \ . \tag{A.28}$$

$$\varepsilon = \varrho s T - P + \mu \varrho + v_k g_k \ . \tag{A.29}$$

We would now like to use the LTE hypothesis to cast the Gibbs equation (A.28) in its local equilibrium form (step 3). This means that we regard all variables (ε, ϱ, s . . . etc.) occurring in (A.28) as local variables, i.e., $\varepsilon = \varepsilon(x, t)$, etc. The differentials $d\varepsilon(x, t)$, etc. now refer to changes involving both x and t. Dividing (A.28) by dt and taking the limit $dt \to 0$, the time derivatives $d\varepsilon/dt$, etc. have to be understood as the material derivatives $D_t \varepsilon(x, t)$, etc. Recall now from (A.5) that this derivative allows one to describe changes corresponding to a fixed particle (or a given local region) of the fluid. Thus the LTE assumption transforms (A.28) into the form

$$D_t \varepsilon - \mu D_t \varrho - v_k D_t g_k = T D_t(\varrho s) \ . \tag{A.30}$$

We now link this equation to the conservation equations (A.10, 13, 22) after we have written down the equation of continuity for the entropy density s.

   We use the fact that the entropy S inside Σ can change either due to exchange of heat or matter with the surroundings, or due to production; this is expressed by the relation

$$D_t S = D_t S^i + D_t S^e \ , \qquad \text{where} \tag{A.31}$$

$$S = \int_V \varrho s \, d^3x \ ; \tag{A.32}$$

the superscript "e" refers to the exchange contribution and "i" refers to internal production. The term $D_t S^e$ can be related to the entropy current J_s:

$$D_t S^e = - \oiint_\Sigma J_s \cdot \hat{n}\, d\Sigma = - \int_V \boldsymbol{V} \cdot J_s\, d^3 x \ , \tag{A.33}$$

where the second equality follows by the Gauss divergence theorem. We may now write

$$D_t S^i = \int_V \sigma_s\, d^3 x \ . \tag{A.34}$$

According to the Second Law, entropy can only be produced, meaning thereby that the function $\sigma_s(x, t)$ must be positive semi-definite:

$$\sigma_s \geqslant 0 \ . \tag{A.35}$$

Combining (A.31–34) and using (A.8, 10) leads to two alternative forms of the entropy conservation equation:

$$\varrho D_t s + \boldsymbol{V} \cdot J_s = \sigma_s \ , \tag{A.36}$$

$$\partial_t(\varrho s) + \boldsymbol{V} \cdot (J_s + \varrho s v) = \sigma_s \ . \tag{A.36'}$$

We now use (A.10, 13, 22, 29, 36) in (A.30) and obtain the result

$$T\sigma_s - T\boldsymbol{V} \cdot J_s = Pv_{k,k} - \sigma_{ik} v_{i,k} - q_{k,k} \ . \tag{A.37}$$

Splitting σ_{ik} into its reversible and dissipative parts, we write

$$\sigma_{ik} = \sigma_{ik}^R + \sigma_{ik}^D \ . \tag{A.38}$$

Then (A.37) can be written as

$$T\sigma_s - T\boldsymbol{V} \cdot J_s = (P\delta_{ik} - \sigma_{ik}^R) v_{i,k} - \sigma_{ik}^D v_{i,k} - q_{k,k} \ . \tag{A.39}$$

Let us choose

$$J_s = q/T \ . \tag{A.40}$$

Then, (A.39) yields the expression

$$\sigma_s = \frac{1}{T}(P\delta_{ik} - \sigma_{ik}^R) v_{i,k} - \frac{1}{T}\sigma_{ik}^D v_{i,k} - \frac{1}{T^2} q_k T_{,k} \ . \tag{A.41}$$

If we ignore dissipation completely, i.e., if we set $\sigma_{ik}^D = 0$ and $q_k = 0$, then there should be no entropy production, i.e., σ_s should vanish. This condition determines

$$\sigma_{ik}^R = P\delta_{ik} \ . \tag{A.42}$$

We then obtain

$$\sigma_s = -\frac{1}{T}\sigma_{ik}^D v_{i,k} + q_k\left(\frac{1}{T}\right)_{,k} . \tag{A.43}$$

There is some arbitrariness in the extraction of J_s and σ_s from (A.39). The choice of J_s and σ_s is dictated by the fact that σ_s should not contain a divergence of a vector field which could change sign at different (x, t) leading to trivial violations of condition (A.35) [A.15]. Also, σ_s must be invariant under a Galilean transformation. Note that the fluxes $\{J_a\}$, q and J_s have to be consistent with the invariance of (A.10, 13, 22, 36) under Galilean transformations [A.7]. To illustrate this point, suppose we had not assumed $g = \varrho v$ as we have done here, and suppose we had started with (A.10) in the lab frame. Then a transformation to the rest frame (denoted by primed symbols) implies

$$x' = x - vt , \qquad t' = t ; \tag{A.44}$$

and

$$\partial_i' = \partial_i , \qquad \partial_t' = \partial_t + v_k\partial_k , \tag{A.45}$$

which transform (A.10) to the equation

$$\partial_t'\varrho + \partial_k'(g_k - \varrho v_k) = 0 . \tag{A.46}$$

Since $\varrho'(x', t') = \varrho(x, t)$ (ϱ being a scalar), we conclude that the combination $g_k = \varrho v_k$ must be identified with g_k' which must vanish by the definition of the rest frame. Thus we derive the mass flux to be $g_k = \varrho v_k$.

The functional forms of $\{J_a^D\}$, the dissipative parts of the fluxes, still remain to be determined. For this purpose, we have to invoke Onsager's theory of linear irreversible thermodynamics [A.4, 5] (recall step 5). Equation (A.43) shows that σ_s can be expressed in the form (A.3). The choice of the flows thus determines the conjugate forces. For example, choosing q as the heat flow determines $V(T^{-1})$ as the conjugate force, and choosing σ_{ik}^D as the (dissipative) momentum flux determines $v_{i,k}/T$ as the corresponding force. The Onsager phenomenological relations connect the fluxes with the appropriate forces via appropriate transport coefficients. In the case of a fluid, these relations are

$$q_k = -\kappa_{kl} T_{,l} ,$$
$$\sigma_{ik}^D = -\eta_{iklm}\tfrac{1}{2}(v_{l,m} + v_{m,l}) , \tag{A.47}$$

where κ_{kl} and η_{iklm} are, respectively, the thermal conductivity and the viscosity coefficients. It should be noted that such linear laws are strictly valid only if the system is sufficiently "close" to equilibrium, where the forces are "small". The overall symmetry of the system is used to determine the number of independent nonvanishing coefficients of the tensors κ_{ik} and η_{iklm}. For the isotropic fluid, there is only one such coefficient for $\{\kappa_{ik}\}$ and two for $\{\eta_{iklm}\}$. Equations (A.47) determine the form of the dissipative fluxes occurring in the hydrodynamic equations.

The hydrodynamic equations (A.10, 13, 22, 36) are not yet closed. There are five equations (noting the redundancy of one of the variables ε, s), but seven variables (ϱ, g_k, s, P, T). The additional variables P and T can be eliminated by using the equation of state (recall step 6). Choosing to work in the (ϱ, s) representation of thermodynamics, one changes variables from (δP, δT) to ($\delta\varrho$, δs):

$$
\begin{pmatrix} \delta P \\ \delta T \end{pmatrix} = \begin{pmatrix} \left(\dfrac{\partial P}{\partial \varrho}\right)_s & \left(\dfrac{\partial P}{\partial s}\right)_\varrho \\ \left(\dfrac{\partial T}{\partial \varrho}\right)_s & \left(\dfrac{\partial T}{\partial s}\right)_\varrho \end{pmatrix} \begin{pmatrix} \delta\varrho \\ \delta s \end{pmatrix} ,
\tag{A.48}
$$

where δP is the deviation of P from the equilibrium value P_0, etc. The matrix appearing on the rhs has to be consistent with the Maxwell relation

$$
\left(\frac{\partial T}{\partial \varrho}\right)_s = \frac{1}{\varrho^2}\left(\frac{\partial P}{\partial s}\right)_\varrho
$$

which follows from the equality

$$
\left(\frac{\partial^2 \varepsilon}{\partial\varrho\,\partial s}\right)_g = \left(\frac{\partial^2 \varepsilon}{\partial s\,\partial\varrho}\right)_g
$$

and the definitions

$$
T = (\partial\varepsilon/\partial s)_{\varrho,g} , \qquad P = \varrho^2(\partial\varepsilon/\partial\varrho)_{s,g} .
$$

Thus, using (A.48), the variables δP and δT can be eliminated from the linearized hydrodynamic equations. The result is a set of five (linearized) equations in five independent variables.

We have now completed the main aim of this Appendix, namely, the setting up of the hydrodynamic equations of a given system, illustrated in detail by the example of a simple fluid. For many applications, these equations [i.e., (A.10, 13, 22)], together with the constitutive equations for the fluxes and the thermodynamic equations of state, are linearized around the equilibrium state of the fluid. These (linearized) equations can be written in a matrix form

$$
\dot\chi_a = -M_{ab}\chi_b .
\tag{A.49}
$$

Some comments about the use of these equations are in order. The main use is the calculation of various correlation functions which are probed by various experiments. The general form of a correlation function involving the variables

χ_a and χ_b is

$$C_{ba}(xt, x't') = V \langle \delta\chi_a(x', t') \delta\chi_b(x, t) \rangle$$
$$= C_{ba}(|x - x'|, |t - t'|) , \tag{A.50}$$

where $\delta\chi_a(x, t) = \chi_a(x, t) - \langle \chi_a(x, t) \rangle$ is the (equilibrium) fluctuation in the variable χ_a and $\langle \ \rangle$ denotes the thermal or ensemble average of $\chi_a(x, t)$. Experiments usually measure the Fourier transform of $C_{ba}(x, t)$:

$$C_{ba}(k, \omega) = \int d^3x \int_{-\infty}^{\infty} dt \, C_{ba}(|x|, t) \exp[i(k \cdot x - \omega t)] . \tag{A.51}$$

Thus the (Rayleigh–Brillouin) light scattering or (coherent) neutron scattering from a fluid measures $S(k, \omega) = C_{\varrho\varrho}(k, \omega)$. All transport coefficients such as η_{iklm}, κ_{ik} etc. are expressible as some correlation function or the other. Thus

$$\kappa_{ik} \sim \int_0^{\infty} dt \, C_{q_i q_k}(t) . \tag{A.52}$$

There are many formalisms for the calculation of correlation functions from hydrodynamic equations—one such formalism is the Landau theory of fluctuating hydrodynamics [A.8]. Though it is important, we do not wish to deal with this aspect here.

As *Martin* et al. [A.1] have noted, one of the simplest applications of (A.49) is to solve the secular equation $\|\tilde{M}_{ab}(k, \omega) - i\omega\delta_{ab}\| = 0$ arising from the Fourier transform of (A.49) ($\tilde{M}$ is the Fourier transform of M). This gives the dispersion relation $\omega = \omega(k)$ of hydrodynamics. The qualitative nature of the solutions of the secular equation is easily inferred from the basic properties of the system. It is easy to specify the number and type of modes that exist for a given system. Martin et al. [A.1] give an exhaustive table listing for various systems the number of parameters entering into the hydrodynamic equations, the number N of independent hydrodynamic variables, the numbers of transport coefficients $\{\eta_{iklm}\}$, $\{\kappa_{ik}\}$, the number of elastic constants, the number of independent thermodynamic derivatives, etc. Thus in a simple fluid, $N = 5$; there are two independent viscosity coefficients and one thermal conductivity coefficient, etc. They also give another exhaustive table (Table II of Ref. [A.1]) listing the number of propagating and diffusive modes in a given system and the number of extra broken-symmetry variables (see Sect. A.3) needed for the hydrodynamic description of that system. Denoting these three numbers by (n_1, n_2, n_3), a simple fluid is characterized by the set $(2, 3, 0)$, whereas a crystal (or a glass) is characterized by the set $(6, 2, 3)$. They also observe that the propagating modes in any given system always occur in pairs (corresponding to the two solutions $\omega = \pm ck$, c being the appropriate propagation velocity) as a result of time-reversal symmetry. On the basis of Onsager's relations (A.4), it is also known [A.17] that the maximum number of propagating modes is twice the number of even or odd hydrodynamic variables, whichever is smaller. [A variable which does not (or does) change its sign under time-reversal is called an even (or odd) variable].

We now describe the generalization of the procedure for setting up the hydrodynamic equations for a system having continuous broken symmetries.

A.2 Ordered Media with Continuous Broken Symmetries

Systems like crystals, liquid crystals, glass, quasicrystals need additional variables—the broken symmetry variables—for their hydrodynamic description. Let these (intensive) variables be denoted by $\xi^\alpha(x, t)$, obeying separate dynamical equations

$$D_t \xi^\alpha = -Z^\alpha \ . \tag{A.53}$$

The Gibbs relation (A.28) generalizes to the form [A.18]

$$d\varepsilon = T d(\varrho s) + \mu \, d\varrho + v_k \, dg_k + h^\alpha d\xi^\alpha + \phi^\alpha_k d\xi^\alpha_{,k} + \psi^\alpha_{kl} d\xi^\alpha_{,kl} \ . \tag{A.54}$$

The thermodynamic "forces" h^α, ϕ^α_k and ψ^α_{kl} must vanish in the limit $k \to 0$ since ξ^α and its gradients are hydrodynamic variables. The Euler relation is same as (A.29). In situations involving extensive broken symmetry variables Ξ^α, one can write

$$\Xi^\alpha(t) = \int_V d^3x \, \xi^\alpha(x, t) \ . \tag{A.55}$$

Then the Gibbs relation (A.28) remains valid whereas the Euler relation becomes

$$\varepsilon = \pi + \mu \varrho + T \varrho s + v_k g_k \ , \quad \text{where} \tag{A.56}$$

$$\pi = -P + h^\alpha \xi^\alpha + \phi^\alpha_k \xi^\alpha_{,k} + \psi^\alpha_{kl} \xi^\alpha_{,kl} \ . \tag{A.57}$$

Thus one can deal with extensive variables by a simple replacement of $-P$ by π. The expression for an entropy source is obtained along the same lines as in a liquid.

A.2.1 Hydrodynamics of a Solid

As an example of a system exhibiting continuous broken symmetries, consider a solid. Translations are described by the Galilean invariant displacement vector R of the lattice sites. In a solid, the (arbitrary) translational invariance in all three directions is spontaneously broken. The three components R_k of the displacement vector can be taken as the broken symmetry variables. These are appended to the five usual hydrodynamic variables $\{\varepsilon, g_k, \varrho\}$ of a liquid. The Gibbs relation appropriate for a solid is

$$Td(\varrho s) = d\varepsilon - \mu d\varrho - v_k dg_k - \phi_{ik} dR_{i,k} \ , \tag{A.58}$$

and the corresponding Euler relation is

$$P - \phi_{ik} R_{i,k} = -\varepsilon + \mu\varrho + T\varrho s + v_k g_k \ . \tag{A.59}$$

(This is appropriate for an extensive variable corresponding to displacement.) Note that here the stress tensor ϕ_{ik} is symmetric; $\phi_{ik} = \phi_{ki}$. Let the dynamical equation for R_k be

$$D_t R_k = Z_k = Z_k^R + Z_k^D \ . \tag{A.60}$$

It is easy to derive the following equation for the deformation gradient $R_{i,k}$:

$$D_t R_{i,k} + v_{l,k} R_{i,l} = Z_{i,k} \ . \tag{A.61}$$

Since R is an even function under time reversal, Z^R must be odd under time reversal [A.1]. The simplest function in an isotropic solid available at hand as a hydrodynamic variable is the velocity v, i.e., it is natural to take

$$Z^R = v \ . \tag{A.62}$$

Then, proceeding as in the example of a liquid, we obtain

$$\sigma_s = -\frac{1}{T} q_k T_{,k} + \phi_{ik} Z_{i,k} - \{\sigma_{ik} - P\delta_{ik} - \phi_{il} R_{l,k}\} v_{i,k} \ . \tag{A.63}$$

The requirement that the reversible parts of fluxes must contribute nothing to σ_s, and the relation $Z^R = v$, together imply that we must have

$$v_{i,k} \{P\delta_{ik} - \phi_{ik} + \phi_{il} R_{l,k} - \sigma_{ik}^R\} = 0 \ , \tag{A.64}$$

which determines σ_{ik}^R:

$$\sigma_{ik}^R = P\delta_{ik} - \phi_{ik} + \phi_{il} R_{l,k} \ . \tag{A.65}$$

Then we obtain

$$\sigma_s = -\frac{1}{T} q_k T_{,k} + \phi_{ik} Z_{i,k}^D - \sigma_{ik}^D v_{i,k} \ . \tag{A.66}$$

It is convenient to modify the second term to the form $-\phi_{ik,k} Z_i^D$ by adding a total divergence term (which vanishes on integration) to σ_s, leading to the result

$$\sigma_s = -\frac{1}{T} q_k T_{,k} - \phi_{ik,k} Z_i^D - \sigma_{ik}^D v_{i,k} \ . \tag{A.67}$$

From this expression one identifies the forces and the fluxes and sets up the phenomenological Onsager relations [A.1]:

$$\sigma_{ik}^{D} = -\eta_{iklm} v_{l,m} \; ,$$

$$q_k = -\kappa_{kl} \delta T_{,l} - \xi_{kl} \phi_{lm,m} \; ,$$

$$Z_k^{D} = -\frac{1}{T} \xi_{lk} \delta T_{,l} - \zeta_{kl} \phi_{lm,m} \; . \tag{A.68}$$

Here the kinetic coefficients η_{iklm}, κ_{kl} etc. have to satisfy appropriate symmetry relations [A.1]. It is important to note that the coefficients ζ_{kl} allow for the possibility of vacancy diffusion in a crystal. We shall return to this point shortly. The closure of the hydrodynamic equations requires setting up the thermodynamic equation of state analogous to (A.48), where the extra components arising from the derivatives of R (or the strain field) enlarges the matrix of derivatives to include the appropriate partial derivatives of strain with respect to ϱ and s [A.1]. Now the substitution of the Onsager relations (A.68), the equations of state and the constitutive relations (A.62, 65) in equations (A.10, 13, 22, 60) leads to the complete set of hydrodynamic equations for a solid.

Since the crystal has five conserved quantities (as in a fluid) associated with the variables $(\varrho, \varepsilon, v_k)$ and three (continuous) broken symmetries described by the variables R_k, there should be eight hydrodynamic modes [A.1]. Earlier treatments (see the review by *Griffin* [A.19]) of the hydrodynamics of a solid have also included R_k as the relevant extra variables. However, it was assumed that the fluctuations in the divergence of R are simply proportional to those of ϱ. In order to see the reason behind this identification, consider (in a crystal) atoms located at $R(l)$ where l is the lattice site index. The instantaneous atomic position is

$$R(l, t) = R(l) + \delta r(l, t) \; , \tag{A.69}$$

where $\delta r(l, t)$ is the displacement of the atom. The displacement density $u(x, t)$ at the point x is related to this microscopic displacement δr by the relation

$$u(x, t) = \sum_l \delta r(l, t) \delta[x - R(l)] \; . \tag{A.70}$$

Assuming each site to be occupied (i.e., no vacancies) leads to the microscopic expression for the number density $n(x, t)$ (which differs from $\varrho(x, t)$ by a mass factor)

$$n(x, t) = \sum_l \delta[x - R(l, t)] \; . \tag{A.71}$$

Now substituting (A.69) in (A.71), making a Taylor expansion in δr and using (A.70), we obtain

$$n(x, t) = n_0 - u_{k,k} = n_0 - R_{k,k} \; , \tag{A.72}$$

where $n_0 = \sum_l \delta[x - R(l)]$ is the average equilibrium number density. We then obtain the relation

$$\delta n = n - n_0 = - R_{k,k} \ . \tag{A.73}$$

However, a real crystal has vacancies. Denoting the vacancy concentration by c, we have

$$c = (n_0 - n)/n_0 \ , \tag{A.74}$$

where n is the actual occupied density in the crystal. Taking the differential of (A.74) we then obtain

$$dc = (n/n_0^2) \, dn_0 - (1/n_0) \, dn \ . \tag{A.75}$$

Now, owing to the fact that it is the fluctuation in the lattice site density (and not the actual density n) which obeys the relation (A.73), we get

$$dc = - (\varrho/mn_0^2) \, dR_{k,k} - (1/mn_0) \, d\varrho \ . \tag{A.76}$$

In order to include vacancies in the hydrodynamics of a solid in a proper manner, the symbol d must be interpreted as in (A.76). Proceeding along these lines, *Fleming* and *Cohen* [A.20] have derived the (linear) hydrodynamic equations of a crystal with vacancies, although the role of vacancies was first recognized by *Martin* et al. [A.1]. It turns out that a crystal has six propagating modes and two diffusive modes—one corresponding to heat diffusion (as in a liquid) and the other associated with diffusion of vacancies. It is a straightforward matter to incorporate vacancies into the nonlinear hydrodynamic formalism of a solid. Incorporating topological defects like dislocations, however, is nontrivial. The next section is devoted to this task.

A.3 The Poisson Bracket Method in Hydrodynamics

In addition to the customary method of deriving hydrodynamic equations outlined in Sect. A.2, there is yet another method—the so-called Poisson bracket (PB) method, based essentially on the fundamental Poisson brackets involving the basic hydrodynamic variables. Being algebraic, it is almost an automatic piece of machinery leading to nonlinear hydrodynamic equations for any complicated system. It turns out that this technique is also well adapted to include topological defects. We shall consider again the example of a crystal, but this time, with dislocations. Prior to that, a brief description of the method is in order [A.6].

We start with the Hamiltonian of the solid:

$$H = \int d^3x \, \varepsilon(g_k, \varrho, s, \beta_l^k) \ , \quad \text{where} \tag{A.77}$$

$$d\varepsilon = v^k \, dg_k + \mu \, d\varrho + T \, d(\varrho s) + \phi_k^l \beta_l^k \tag{A.78}$$

with the definition $\beta_l^k \equiv R^k{}_{,l}$, see (A.58). We note that g_k is a covariant vector (density) whereas the position (or displacement) R^k (or u^k) is a contravariant vector density.

If we know the Poisson brackets involving the fundamental variables g_k, ϱ, s, β_k^i (these span the "phase space" of hydrodynamics), then we can set up the equations of motion

$$\partial_t a_m + \{a_m, H\} = 0 , \quad (a_m = g_k, \varrho, \dots) . \tag{A.79}$$

Note that the Poisson brackets between hydrodynamic variables are universal since they depend only on the symmetry of the physical laws, and that they are *independent* of the form of the Hamiltonian, H. The equations of motion (A.79) will not, of course, include the effects of dissipation, which has to be appended phenomenologically to these "reversible equations" by means of the dissipation function $\mathcal{R}$. Then (A.79) would read

$$\partial_t a_m + \{a_m, H\} = -\partial \mathcal{R}/\partial \tilde{a}_m , \tag{A.80}$$

where $\tilde{a}_m$ is the variable thermodynamically conjugate to a_m. Note that the function $\mathcal{R}$ occurs in the expression for σ_s:

$$\sigma_s T = \sum_m (\partial \mathcal{R}/\partial \tilde{a}_{m,k}) \tilde{a}_{m,k} . \tag{A.81}$$

Thus the Poisson bracket method is really meant to derive the dissipation-free reversible (but nonlinear) hydrodynamic equations.

Let $\{l_\alpha | \alpha = 1, \dots, N\}$ be the generators of the underlying hydrodynamic group G. These belong to the Lie algebra $\mathcal{G}$ of G. (Appendix E.) The variables $\{a_m\}$ together with $\{l_\alpha\}$ form another Lie algebra. An infinitesimal change at a point x_1 in the variable a_m is given by

$$\delta a_m(x_1) = \int \{l_\alpha(x_2), a_m(x_1)\} \delta\theta_\alpha(x_2) dx_2 , \tag{A.82}$$

where $\theta_\alpha(x)$ is the group parameter corresponding to $l_\alpha(x)$. Note here that the Lie bracket, or the composition law in the Lie algebra, is the PB. From (A.82) we obtain the expression for the PB in terms of the *variational* derivatives:

$$\{l_\beta(x_2), l_\alpha(x_1)\} = \frac{\delta l_\alpha(x_1)}{\delta \theta_\beta(x_2)} . \tag{A.83a}$$

Similarly, one obtains

$$\{l_\beta(x_2), a_m(x_1)\} = \frac{\delta a_m(x_1)}{\delta \theta_\beta(x_2)} . \tag{A.83b}$$

However, the $\{a_m(x)\}$ satisfy the relations

$$\{a_k(x_1), a_m(x_2)\} = 0 . \tag{A.83c}$$

For the case of solids, the group G is the translation group, the generator being the total momentum; and the parameter, the displacements $u^k(x)$: $x^k \mapsto x^k + u^k(x)$. The variational derivatives for ϱ, s and R_k are

$$\frac{\delta \varrho_2}{\delta u_1^k} = \{g_{k1}, \varrho_2\} , \qquad (A.84a)$$

$$\frac{\delta s}{\delta u_1^k} = \{g_{k1}, s\} , \qquad (A.84b)$$

$$\frac{\delta R_2^l}{\delta u_1^k} = \{g_{k1}, R_2^l\} , \qquad (A.84c)$$

where we have used the abbreviation $\varrho_2 \equiv \varrho(x_2)$, $g_{k1} = g_k(x_1)$ etc. Now $\varrho(x)$ and $s(x)$ are scalar densities of weight 1 (since $\int \varrho(x) d^3x$ is a scalar and $d^3x \to |\partial x_i/\partial x_k'| d^3 x'$ under the transformation $x^k \to x'^k = x^k + u^k$, $\varrho(x) \to \varrho(x')|\partial x_i'/\partial x_k|$ etc.).
Hence

$$\varrho'(x') = \left| \frac{\partial x_i}{\partial x_k'} \right| \varrho(x) = (1 - u_{l,l}) \varrho(x) + O(u^2) ,$$

and therefore

$$\begin{aligned}
\delta\varrho(x) &= \varrho'(x) - \varrho(x) \\
&= \varrho'(x') - \varrho(x) + \varrho'(x) - \varrho'(x') \\
&= -\varrho(x)u_{l,l} + \varrho'(x) - \varrho'(x+u) \\
&= -\varrho(x)u_{l,l} - u^l \partial_l \varrho'(x) \\
&= -\varrho(x)u_{l,l} - u^l \partial_l \varrho(x) + O(u^2) .
\end{aligned}$$

Thus,

$$\begin{aligned}
\frac{\delta\varrho_2}{\partial u_1^k} &= -\varrho_2 \partial_{12}(\delta u_2^l/\delta u_1^k) - (\delta u_2^l/\delta u_1^k)\partial_{12}\varrho_2 \\
&= -\varrho_2 \partial_{k2}\delta(1-2) - \delta(1-2)\partial_{k2}\varrho_2 \\
&= -\partial_{k2}[\varrho_2\delta(1-2)] = +\varrho_1 \partial_{k1}\delta(1-2) . \quad [\delta(1-2) \equiv \delta(x_1 - x_2), \text{ etc.}]
\end{aligned}$$

Comparing this with (A.84a) we obtain

$$\{g_{k1}, \varrho_2\} = \varrho_1 \partial_{k1} \delta(1-2) . \qquad (A.85a)$$

Similar reasoning leads to the result

$$\{g_{k1}, s_2\} = s_1 \partial_{k1} \delta(1-2) \ . \tag{A.85b}$$

The transformation law for g_k (a covariant tensor density of weight 1) is

$$\delta g_k = -u^l g_{k,l} - g_l u^l_{,k} - g_k u^l_{,l} \ .$$

Then a straightforward calculation gives

$$\{g_{k1}, g_{k2}\} = \delta g_{l2}/\delta u^k_1 = g_{l1} \partial_{k1} \delta(1-2) - g_{k2} \partial_{l2} \delta(1-2) \ . \tag{A.85c}$$

Similarly, using the relation

$$\delta \beta^l_k = -u^m \beta^l_{k,m} - \beta^l_m u^m_{,k} \ ,$$

we obtain the PB

$$\{g_{k1}, \beta^m_{l2}\} = \delta \beta^m_{l2}/\delta u^k_1 = -\delta(1-2) \beta^m_{l2,k1} - \beta^m_{k2} \partial_{l2}(1-2) \ . \tag{A.85d}$$

All the other PB's work out to zero.

The derivation of the reversible equation now becomes elementary. Thus, using the PB (A.85a),

$$\begin{aligned}
\partial_t \varrho = \{H, \varrho\} &= \int dx_1 v^k_1 \{g_{k1}, \varrho\} \\
&= \int dx_1 v^k_1 \varrho_1 \partial_{k1} \delta(x_1 - x) = -\partial_k(\varrho v^k) \ ,
\end{aligned} \tag{A.86}$$

which coincides with the earlier result, (A.10). The momentum equation is

$$\partial_t g_k + (P\delta^l_k + g_k v^l + \beta^m_k \phi^l_m)_{,l} = 0 \ , \tag{A.87}$$

where

$$P = -\varepsilon + g_k v^k + T\varrho s + \varrho \mu \ .$$

When dislocations are present in the solid, the functions β^l_k do not satisfy the integrability conditions $\beta^m_{l,k} - \beta^m_{k,l} = 0$, and one naturally introduces the non-integrable part

$$\alpha^m_{kl} = \beta^m_{l,k} - \beta^m_{k,l} \tag{A.88}$$

as the dislocation–density tensor. (Sometimes, the dual tensor α^n_m defined as $\alpha^n_{kl} = \alpha^n_m \varepsilon_{klm}$ is called the dislocation–density tensor). A straightforward calculation using PB's then leads to the equation

$$\partial_t \beta^m_k = \{H, \beta^m_k\} = -v^l \beta^m_{k,l} + \beta^m_l v^l_{,k} \ .$$

This equation can be written in the more useful form

$$\partial_t \beta^m_k + (v^l \beta^m_l)_{,k} - v^l \alpha^m_{kl} = 0 \ . \tag{A.89}$$

Now let the dislocation flux be defined as

$$J_k^m = \partial_t \beta_k^m + (v^l \beta_l^m)_{,k} \; . \tag{A.90}$$

Then (A.89) implies the following conservation law:

$$\partial_t \alpha_{kl}^m + J_{k,l}^m - J_{l,k}^m = 0 \; . \tag{A.91}$$

Note that in the absence of dissipation

$$J_k^m = v^l \alpha_{kl}^m \; , \tag{A.92}$$

i.e., dislocations are driven away with the velocity of the medium. When dissipation is included and the actual mechanism of entropy production due to dislocations is taken into account (this has to be done phenomenologically), the dislocation conservation equation (A.91) gets coupled to the other hydrodynamic variables of the solid. To our knowledge, such a theory has not been developed so far in full detail.

A.4 Summary

In this Appendix we have given a detailed derivation of the conventional hydrodynamic equations of a simple fluid, assuming that the first and the second laws of thermodynamics remain valid locally, even though the fluid is not in (global) thermodynamic equilibrium. All dissipative effects are included in the expression for the entropy production rate, which is a sum of various terms, each one being bilinear in a thermodynamic force and the corresponding flux; the Onsager law relates the forces to the fluxes by means of the transport coefficients. The hydrodynamic equations are equations of continuity for the density, the momenta, the energy and the entropy. These equations form a closed set when appropriate constitutive laws and the thermodynamic equations of state are used. This is the classical approach to the study of hydrodynamics in condensed matter physics. By considering the example of a crystalline solid, we have indicated how simple hydrodynamics can be extended to the case in which broken symmetries occur. We have also discussed an alternative method of setting up hydrodynamics using the algebraic machinery of Poisson brackets, and indicated how the formalism could be applied to solids with dislocations. For the use of this method in the case of nematic and cholesteric liquid crystals, we refer to the work of *Volovik* and *Kats* [A.21]. A few other applications in condensed matter physics are given in the work of *Dzyaloshinskii* and *Volovik* [A.6]. The Poisson bracket method is currently being employed extensively in setting up the hydrodynamics of diverse systems in other areas of physics such as relativistic fluid dynamics [A.22], plasmas in astrophysical contexts [A.23], etc.

B. Curved Space and Parallel Transport of Vectors

In several places in the text, we have been concerned with curved spaces. It is helpful to recapitulate some of their properties [B.1–3].

Consider a general surface Σ in E^3. Σ is described as the locus of the vector r parametrized by two real variables u^k $(k = 1, 2)$:

$$r = r(u^1, u^2)$$
$$= [x(u^1, u^2), y(u^1, u^2), z(u^1, u^2)] \ , \tag{B.1}$$

where x, y, z are the Cartesian components of r. A curve C lying on Σ is represented by the locus of $r[u^1(\lambda), u^2(\lambda)]$, where λ is an additional real parameter. The vector tangent to C at a point r is given by

$$\frac{dr}{d\lambda} = \frac{dr}{du^k} \cdot \frac{du^k}{d\lambda} \equiv x^k \frac{du^k}{d\lambda} \tag{B.2}$$

[summation over k $(= 1, 2)$ is understood]. The square of a line element of C is given by the relation

$$dl^2 = dr \cdot dr = x_i \cdot x_k \, du^i \, du^k$$
$$= g_{ik} \, du^i \, du^k \ , \tag{B.3}$$

where $g_{ik} = x_i \cdot x_k (= g_{ki})$ is known as the *metric* tensor on Σ. This metric belongs to a special type of space known as a *Riemannian space*. In a Riemannian space, the metric is such that $g_{ik} u^i u^k > 0$ for all $u \neq 0$, and $\|g_{ik}\| \neq 0$. Thus the surface of a sphere of radius R, with its centre at the origin of coordinates in E^3, is a Riemannian space. The length element on the sphere is given by

$$dl^2 = (R \, d\theta)^2 + (R \sin \theta \, d\varphi)^2 \ . \tag{B.4}$$

Here $u^1 = \theta$, $u^2 = \varphi$, $g_{11} = R^2$, $g_{22} = R^2 \sin^2 \theta$, $g_{12} = g_{21} = 0$.

B.1 Parallel Transport of Vectors

In Euclidean space, a global Cartesian system of coordinates can be defined. This can be "*parallel transported*" (i.e., the unit vectors of the basis can be moved parallel to themselves) freely to all points in the space, and the relationships between the coordinate systems at different points can then be written down easily. It is not possible to generalize this in a trivial manner to vectors in a curved space. The result of "parallel transport" of vectors in a curved space depends on the actual path traversed. Moreover, the notion of parallel transport is required to compare the character of the geometry at different points in the continuum. Levi–Civita introduced a notion of parallelism that gives the most effective solution to the problem. Although this was originally proposed for curved surfaces embedded in a Euclidean space, it has been generalized to an arbitrary Riemannian geometry on a manifold.

Let us consider again the case of a 2D (curved) surface Σ in E^3. Vectors in Σ may be defined most naturally by choosing them to lie in the tangent plane at each point. Consider a vector v_P at a point $P \in \Sigma$ (Fig. B.1). Let $Q \in \Sigma$ be a point close to P such that the angle between the tangent planes T_P and T_Q (at P and Q, respectively) is an infinitesimal, δ. Our aim is to construct a vector v_Q at Q that is parallel to v_P. (It is to be remembered that only Σ "exists", the 3-space in which Σ is embedded being an artifact to be removed subsequently.) We move v_P parallel to itself in E^3 such that its tail is at Q. Call this shifted vector v_P'. Let v_Q be the orthogonal projection of v_P' onto T_Q. v_Q makes the smallest angle with v_P', and is as nearly parallel to v_P', as is compatible with its lying in T_Q. One may then accept v_Q as the desired "parallel" vector at Q.

v_P and v_Q differ in angle by a first order infinitesimal (δ) and in magnitude by $O(\delta^2)$. $(v_Q - v_P)$ is normal to Σ to within second order terms. A set of vectors $\{v_P^{(i)}\}$ at P goes over into a set $\{v_Q^{(i)}\}$ at Q under this procedure, preserving the angles between the members of the set. However, this prescription for parallel transport over small regions fails to give a unique vector field when done in the large. To extend the notion of parallelism to a whole region in a meaningful way, Levi–Civita introduced the idea of parallel transport *along a given curve*. It turns out that this is the best that can be done in curved space; in fact, this idea turns out to be the correct fundamental concept. Let the scalar variable λ parametrize a curve C on Σ, and let P, Q be nearby points on C. In the limit $Q \to P$, $dv/d\lambda = \lim_{\Delta\lambda \to 0} (\Delta v/\Delta\lambda)$, where $\Delta v = (v_Q - v_P')$ is normal to Σ. We may therefore *define* $dv/d\lambda$ along the curve C such that it is always normal to Σ. Let X_1, X_2 be the components of the tangent vector to C at P, along linearly independent

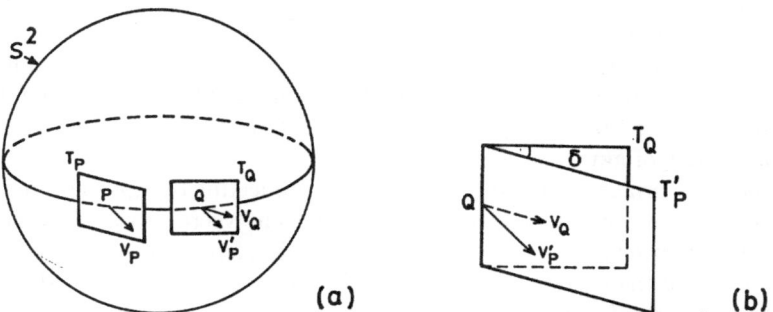

Fig. B.1a, b. Approximate local parallelism in the sense of Levi–Civita. The surface is taken to be S^2. (a) the vector v_P lies in T_P, but the vector v_P', which is parallel to v_P at Q (in the surrounding space E^3) does not lie in T_Q. The orthogonal projection of v_P' on T_Q is v_Q. (b) intersection of T_Q with a plane T_P' which is parallel to T_P; the angle between T_Q and T_P' is δ. Here $v_P' \in T_P'$ and $v_Q \in T_Q$

directions in T_P. Then the pair of equations

$$\frac{dv}{d\lambda} \cdot X_1 = 0 \, , \qquad \frac{dv}{d\lambda} \cdot X_2 = 0 \qquad\qquad\qquad (B.5)$$

determines $v(\lambda)$ uniquely, given $v(\lambda_0)$ at some initial point on C. Parallelism along a *given* curve on Σ, as defined by (B.5), has the following properties: (a) uniqueness of the vector resulting from parallel transport; (b) invariance of $(v \cdot w)$ along C for any pair of vectors v and w; (c) constancy of the length of a vector $v(\lambda)$ along C; (d) congruency of any bundle of vectors to itself in parallel transport along C; (e) reduction to the familiar form of parallelism in Euclidean space. Finally, if the result of parallel transport is independent of the actual path traversed at all points in a space, the space is Euclidean; the converse is also true.

The *Gaussian curvature* κ of a surface (which turns out to be an intrinsic property of the surface) has a fundamental connection with parallel transport. Let a vector v be parallel transported around a smooth closed curve C on Σ. Suppose C surrounds a point P, and bounds an infinitesimal area ΔA. (We adopt the usual convention of keeping the area on our left as we traverse C.) In general, the vector v' obtained on parallel transporting v (from any starting point) once around C will not coincide with the original v. There will be an angle of mismatch $\Delta\alpha$ between v' and v. The Gaussian curvature at P is defined as

$$\kappa(P) = \lim_{\Delta A \to 0} (\Delta\alpha/\Delta A) \, , \qquad\qquad\qquad (B.6)$$

with the proper sign for $\Delta\alpha$. For a curve bounding a finite, simply-connected domain of Σ, the integral of κ over the area enclosed by the curve is equal to the angle of mismatch α arising from the parallel transport of a vector around the curve. This is the essential content of the well-known *Gauss–Bonnet theorem*. If $\kappa = 0$ everywhere, α vanishes. The converse is also true. Surfaces in 3-space with this property are "*developable*". They can be mapped isometrically on the plane, i.e., the mapping preserves distances between points. S^2 is not developable, whereas the surface of a cylinder is. Surfaces with *constant* Gaussian curvature have the important property that one can move tiles on them without changing the shape of the tiles. This capacity for rigid motion explains why it is meaningful to talk of tiling S^2 or S^3 ($\kappa > 0$), E^2 or E^3 ($\kappa = 0$), and H^2 or H^3 ($\kappa < 0$) as, for instance, in theories of amorphous structures.

A *geodesic* on a surface Σ is the shortest arc on Σ connecting two given points. On S^2, geodesics are arcs of great circles. The tangent vector to a geodesic remains so when parallel transported along the geodesic. Hence geodesics can be defined as curves on the surface whose tangent vectors form a parallel field—a natural generalization of straight lines in Euclidean spaces.

On E^2, the circumference-to-radius ratio of a circle of radius ϱ is $L/\varrho = 2\pi$. On the surface S^2 of a sphere of radius R (Gaussian curvature $\kappa = 1/R^2$),

$$\frac{L}{\varrho} = 2\pi \frac{\sin(\varrho\kappa^{1/2})}{\varrho\kappa^{1/2}} < 2\pi \, . \qquad\qquad\qquad (B.7)$$

On a hyperbolic surface H^2 of constant curvature $\kappa < 0$,

$$\frac{L}{\varrho} = 2\pi \frac{\sinh(\varrho|\kappa|^{1/2})}{\varrho|\kappa|^{1/2}} > 2\pi \ . \tag{B.8}$$

In each case, L and ϱ are of course the corresponding geodesic lengths.

B.2 The Covariant Derivative

We will now examine how derivatives of vector fields on curved surfaces may be defined. Consider a vector field $v = \{v^k\}$ in a space of n dimensions. The index k ($= 1, 2, \ldots, n$) labels the components of the vector: $v^k = v^k(x^1, \ldots, x^n)$. The change Δv^k when v is parallel transported from the point x to the point $(x + \Delta x)$ can be written in general as the sum of two terms: the first comes from the explicit dependence of v^k on the coordinates; the second term arises from the difference in the geometry as one moves from x to $x + \Delta x$. It is reasonable to expect that this term will be linear in Δx, and also proportional to the value of the vector field in the neighbourhood concerned. We may thus write

$$\Delta v^k = \frac{\partial v^k}{\partial x^m} \Delta x^m + \Gamma^k_{lm} v^l \Delta x^m \ , \tag{B.9}$$

where a summation is implied over repeated indices. The coefficient Γ^k_{lm} is called a Christoffel symbol or *connection*. From (B.9),

$$\lim_{\Delta x^m \to 0} \frac{\Delta v^k}{\Delta x^m} \equiv \frac{Dv^k}{Dx^m} = \frac{\partial v^k}{\partial x^m} + \Gamma^k_{lm} v^l \ . \tag{B.10a}$$

Dv^k/Dx^m is called the *covariant derivative* of the vector field. In more compact notation

$$D_m v^k = \partial_m v^k + \Gamma^k_{lm} v^l \ . \tag{B.10b}$$

The above definition may now be used to quantify parallel transport by imposing the condition of orthogonal projection referred to earlier. Let $x^i(\lambda)$ represent the path (on a 2D curved surface Σ) along which v is parallel transported, and let u^l be the tangent vector ($u^1 = dx^1/d\lambda$, $u^2 = dx^2/d\lambda$) to the path. The perpendicularity condition implies that $u^m D_m v^k = 0$, which reduces to

$$\frac{dv^k}{d\lambda} + \Gamma^k_{lm} v^l u^m = 0 \ . \tag{B.11}$$

In terms of infinitesimal changes, this says that $\delta v^k = -\Gamma^k_{lm} v^l \delta x^m$ is the change in the vector field owing to the change in geometry alone. It is this effect that is compensated for in the definition of the covariant derivative.

The covariant derivative of a scalar (i.e., single-component) function coincides with its usual derivative. The covariant derivative of a covariant vector field

v_k (preferably called a covector) can be shown to be

$$D_m v_k = \partial_m v_k - \Gamma^l_{km} v_l \ . \tag{B.12}$$

Using (B.10) and (B.12), one can find the covariant derivatives of tensors of higher rank.

The concept of a connection is actually independent of that of a metric. Within the framework of Riemannian geometry, the connection is said to be compatible with the metric if $D_m g_{ij} = 0$. If we restrict ourselves further to symmetric connections $\Gamma^k_{lm} = \Gamma^k_{ml}$, then the metric determines the connection according to

$$\Gamma^k_{lm} = \tfrac{1}{2} g^{kj} (\partial_l g_{jm} + \partial_m g_{lj} - \partial_j g_{lm}) \ . \tag{B.13}$$

In our example of S^2 with coordinates $x^1 = \theta$, $x^2 = \varphi$, the nonzero Γ^k_{lm} are $\Gamma^1_{22} = -\sin\theta\cos\theta$, $\Gamma^2_{12} = \Gamma^2_{21} = \cot\theta$.

There exists yet another type of connection (the *internal connection*), which is not derivable from a metric and which is of fundamental significance in gauge theories (see Appendix F).

B.3 The Curvature

Let us now consider the parallel transport of a vector $\boldsymbol{n}$ around an infinitesimal loop made up by placing the vectors $\boldsymbol{a}$, $\boldsymbol{b}$, $-\boldsymbol{a}$ and $-\boldsymbol{b}$ tail to head. As we have seen, the final vector $\boldsymbol{n}'$ will not coincide with the initial vector $\boldsymbol{n}$ if the surface is curved. It turns out that $\boldsymbol{n}'$ can be written in the form

$$n'^i = n^i - R^i_{jkl} n^j a^k b^l \ , \quad \text{where} \tag{B.14a}$$

$$R^i_{jkl} = \partial_k \Gamma^i_{jl} - \partial_l \Gamma^i_{jk} + \Gamma^i_{mk} \Gamma^m_{jl} - \Gamma^i_{ml} \Gamma^m_{jk} \tag{B.14b}$$

is the Riemann–Christoffel tensor. In an n-dimensional space, this tensor has, as a result of its symmetries, $n^2(n^2 - 1)/12$ independent components. The curvature tensor is the contraction $R_{ij} = R^k_{ijk}$. The scalar curvature is $\mathscr{R} = g^{ij} R_{ij}$. Finally, the Gaussian curvature is

$$\kappa = \mathscr{R}/2 \ . \tag{B.15}$$

If S^2 is a sphere of radius R in E^3, then $\kappa = 1/R^2$, as already stated. The Gaussian curvature of a surface remains invariant under any deformation of the surface that preserves the lengths of curves on it.

B.4 The Torsion

On a flat surface, a parallelogram may be constructed as follows. Let $\overrightarrow{OA} = \boldsymbol{a}$ and $\overrightarrow{OB} = \boldsymbol{b}$ be nonparallel vectors. Parallel transport $\boldsymbol{a}$ along OB to a final position $\overrightarrow{BC}$, and $\boldsymbol{b}$ along OA to a final position $\overrightarrow{AC'}$. As the surface is flat, C and C'

coincide to yield a parallelogram $OACB$. On a general curved surface, this happens if and only if the connection Γ^k_{lm} is symmetric, i.e., $\Gamma^k_{lm} = \Gamma^k_{ml}$. The difference

$$\Gamma^k_{lm} - \Gamma^k_{ml} = T^k_{lm} \tag{B.16}$$

is a tensor, called the *torsion*. It is a measure of the *gap* between the tips of infinitesimal vectors a and b after parallel transport in the manner described above. We recall that the curvature is deduced from the *angle* of mismatch after parallel transporting a vector around an infinitesimal circuit. For this reason, torsion and curvature occur naturally in continuum theories of defects in condensed matter, in connection with translational (dislocation) and rotational (disclination) defect densities [B.4].

Unlike the usual partial derivatives ∂_i, ∂_k, etc., the covariant derivatives D_i and D_k do not commute with each other in general. The commutator of D_i and D_k is, in fact, directly related to the curvature and the torsion (i.e. the defect densities, in the physical context referred to above). One finds

$$[D_i, D_k] v^l = D_i D_k v^l - D_k D_i v^l = R^l_{mik} v^m + T^m_{ik} D_m v^l \ . \tag{B.17}$$

For a symmetric connection, $[D_i, D_k]$ thus provides direct information on the curvature tensor. This fact is of fundamental importance in gauge field theories, in which the analogue of the curvature tensor is the field strength.

B.5 Mapping from Curved Space to Flat Space

We have considered in Chap. 5 instances of the mapping of structures from curved spaces (e.g., S^2 and S^3) to corresponding flat or Euclidean spaces. As pointed out there, such a mapping can be done in many ways [B.5]. Conformal mapping preserves angles but changes lengths by large amounts. Geodesic mapping maps geodesics to straight lines, but alters lengths even more severely. Isometric mapping preserves distances between atoms (bond lengths), and is therefore the most appropriate procedure; but this cannot be done globally, because spaces such as S^2, S^3, H^2, H^3, etc. are not developable. ("Developable" also means complete isometry.) However, parallel transport along a given path in the curved manifold *is* a linear isometric operation. This means that *along a given (directed) path* in the curved manifold, one can *roll* the curved manifold, without sliding, on a Euclidean manifold—just as a sphere can be rolled on a plane along any path drawn on the surface of the sphere to yield an image or tracing of the path on the plane. Clearly, distances between points are unaltered in this process of tracing by rolling the sphere. If a vector is attached to any point of the path, making a given angle with the tangent to the path at that point, this angle is unaltered by the mapping procedure. Suppose the path is an infinitesimal circle in the curved manifold, enclosing an area ΔA, and the rolling is done from a point P on the path until one returns to P. Then, (i) a bundle of vectors attached to P is rotated (at the end of the rolling) relative to its initial orientation by an angle $\kappa \Delta A$ (B.6); and (ii) the image path is in general not a closed one, although the displacement of the fiducial point P is a second-order infinitesimal.

This technique of implementing parallel transport by rolling the curved manifold on a Euclidean one is equivalent to defining a covariant derivative. This is also precisely the method used in Sect. 6.5 in a one-dimensional example (recall also Fig. 6.5). The technique has been used in the literature to map from S^3 and H^3 to E^3 in regard to DRP structures [B.5, 6].

C. DRP Structures, Polytopes and the Tiling of S^3

In Sects. 5.1 to 12, we discussed various aspects of a formalism that seeks to describe amorphous structures (in particular, the dense random packing of spheres) with the help of regular tilings of a curved space (specifically, S^3). Here we make some additional remarks to elaborate upon certain points made in the text.

The basic "unit" in the random packing of equal-sized atoms (hard spheres of radius a) is the regular tetrahedron of side $2a$ with the centres of the atoms at the vertices. As stated in Sect. 5.3, E^3 cannot be tiled by these tetrahedra without gaps. Five of them can be fitted around a common edge, but a gap of about 7.5° is left unfilled. If a slight distortion of these tetrahedra (i.e., of bond lengths) is permitted, this gap can be closed. Twenty such tetrahedra can be made to meet at a central vertex, forming an icosahedral cluster of 13 atoms, the central atom being in contact with the 12 outer atoms. (We recall from Sect. 5.1 that the icosahedron is the Platonic solid $\{3, 5\}$.) E^3 cannot be tiled by regular icosahedra, either. The best that can be done is a quasiperiodic tiling that has icosahedral orientational LRO, as stated in the text.

In S^3, tetrahedra *can* be packed five around an edge without any gap. [In the simpler S^2 example, 5 equilateral triangles can be packed around a common vertex without leaving a gap; the circumference-to-radius ratio $L/\varrho < 2\pi$ in this case, as in (B.7).] Suppose the Gaussian curvature of the space S^3 is κ (i.e., the radius of the hypersphere in E^4 is $\kappa^{-1/2}$). We imagine an icosahedral cluster of atoms (hard spheres of radius a) to be imbedded in S^3. Let $2d$ be the distance between the centres of adjacent surface atoms in the cluster; $2a$ is the distance between the centre of a surface atom and the centre of the atom in the middle. Then d and a are related [C.1] according to

$$(\tau + \tau^{-1})\cos(2\kappa^{1/2}d) = \tau + \tau^{-1}\cos(4\kappa^{1/2}a) , \tag{C.1}$$

where $\tau = (1 + \sqrt{5})/2$ is the golden ratio. In the limit $\kappa \to 0$ (Euclidean space), this leads to $d/a \simeq 1.0515$ (recall also Figs. 6.3a and b). On the other hand, one obtains a *regular* tessellation with $d = a$ for a value $\kappa = \kappa^*$ of the curvature, where

$$\kappa^{*1/2} = (1/2a)\cos^{-1}(\tau/2) \ . \tag{C.2}$$

The radius of the hypersphere whose surface is S^3 is thus fixed, for a given a, at the value

$$\kappa^{*-1/2} \simeq 3.183a \ . \tag{C.3}$$

This regular tessellation corresponds, in fact, to the regular polytope $\{3, 3, 5\}$ in E^4. As explained in the text (Table 5.1), it consists of 600 tetrahedral cells, 120 vertices (atoms), and 720 pentacoordinated edges. Although it might appear that $3.183a$ is a rather small value of the radius of S^3, considering that 120 atoms of radius a have to be accommodated in it, we must remember that we have an extra dimension to reckon with. The hypersurface S^3 is a volume $2\pi^2\kappa^{*-3/2}$; the volume of 120 spheres of radius a turns out to be only about 80% of this volume.

When this perfect tiling of S^3 is flattened out so as to exist in E^3, cuts have to be made, and extra material (tetrahedra) inserted in the gaps. The edges of the cuts become line defects (here, disclinations). This inevitable occurrence of topological defects can in fact be turned to advantage by the systematic iterative procedure [C.2] described in Sect. 5.8. A positive feature of this disclination procedure is the incorporation of a rapidly increasing number of atoms in the structure at each successive stage of the iteration.

A remark is in order here on why $\{3, 3, 5\}$ is preferred as a starting point for the DRP structure over the two other regular tilings of S^3 that are also made up of tetrahedral units (see (5.3)). These are $\{3, 3, 3\}$ and $\{3, 3, 4\}$. The reason is that they have lower packing densities and coordination numbers ($Z=4$ and 6, respectively) than $\{3, 3, 5\}$ (which has $Z = 12$). The (flat space) average value of Z for a DRP structure is far closer to the ideal icosahedral value of 12 than it is to 4 or 6.

Finally, in Sect. 5.11 we commented briefly on the possibility of using regular tilings of the hyperbolic space H^3 as a starting point for generating amorphous structures in E^3. Of the eight possible regular polytopes in this case, only $\{3, 5, 3\}$ is made up of icosahedra. (None of them is made up of tetrahedra.) The advantage of starting with a regular tiling of H^3 is that an infinite number of atoms can be accommodated right away, in contrast to the case of compact spherical spaces. Now, the coordination number Z corresponding to the tessellation $\{p, q, r\}$ can be shown to be [C.3]

$$Z = 4q/(2r + 2q - qr) \ . \tag{C.4}$$

(For $q = 3$, this reduces to the familiar formula $Z = 12/(6 - r)$ used commonly in discussing random packings of spheres.) When $q = 5$, we get $Z = 20/(10 - 3r)$. For $\{3, 5, 3\}$ this yields $Z = 20$, rather too large a value, as mentioned in Sect. 5.11.

D. Some Aspects of Group Theory

As mentioned in the preface, we suppose that the reader has a working knowledge of group theory. However, for a proper understanding of symmetry breaking, homotopic classification of defects, and so on, one needs to be familiar with concepts such as group action, orbits of a group, little groups, etc. What follows is a primer on such topics; further details can be found in references [D.1–3].

D.1 Group Morphisms

Given two groups G and G', a map f from G to G', denoted $f: G \to G'$ or $G \overset{f}{\to} G'$ is called a *group morphism* if it preserves the group structure. As an example, consider the cyclic group C_2 and the group $\mathbb{Z}_2$ of integers modulo 2. These groups are given by the multiplication tables

.	e	a
e	e	a
a	a	e

+	0	1
0	0	1
1	1	0

The map $f: C_2 \to \mathbb{Z}_2$, defined by $f(e) = 0$ and $f(a) = 1$, is structure-preserving since $f(e.a) = f(a) = 1 = 0 + 1 = f(e) + f(a)$, etc. This map being one-to-one and onto[1], is called an *isomorphism*. If instead, the map f is a many-to-one structure-preserving structure, it is called a *homomorphism*. An isomorphic map from a group G to itself is called an *automorphism*. As an example, consider the group $G = \mathbb{Z}_3 = \{0, 1, 2\}$ with addition of integers modulo 3 as the composition law. The multiplication table for $\mathbb{Z}_3$ is as below.

+	0	1	2
0	0	1	2
1	1	2	0
2	2	0	1

The only two possible automorphisms of $\mathbb{Z}_3$ are:

$$f: \begin{array}{c} 0 \mapsto 0 \\ 1 \mapsto 1 \\ 2 \mapsto 2 \end{array} \qquad g: \begin{array}{c} 0 \mapsto 0 \\ 1 \mapsto 2 \\ 2 \mapsto 1. \end{array}$$

---

[1] A function $f: G \to G'$ is said to be *one-to-one* if distinct elements in G have distinct images. A function $f: G \to G'$ is said to be *onto* if every $g' \in G'$ is the image of some $g \in G$.

The set of automorphisms $\{f, g\}$ is written Aut $\mathbb{Z}_3$. On composing the maps f and g, we observe that $f \circ f = f, f \circ g = g \circ f = g$ and $g \circ g = f$, which is precisely the composition law for the group $\mathbb{Z}_2$. Thus the (set of) automorphisms of $\mathbb{Z}_3$ themselves form a group, $\mathbb{Z}_2$; one writes $\text{Aut}(\mathbb{Z}_3) = \mathbb{Z}_2$. This is a general property: the set of automorphisms of a group G is a group under the composition of the maps describing them. In a group G, for a fixed $g \in G$, the map $i_g : x \mapsto g$ $xg^{-1}(x \in G)$ is an automorphism, called an *inner automorphism* (or *conjugation* by g).

D.2 Transformation Group, Group Action and Orbits

Inspection of the multiplication table of any finite group shows that every row is merely a permutation of the first row; further, every row is obtained by multiplying any given row (say the first) by a certain element of the group. This means that every finite group G (of order n, say) can be described in terms of the permutation of n elements. The permutations of n objects form a group (denoted by S_n) consisting of $n!$ elements; G is therefore isomorphic to a subgroup of S_n.

Any subgroup T of the group S(X) of all permutations of the elements of a set X is called a *transformation group* on X. As an example, consider three objects 1, 2 and 3 constituting a set X. The group $S_3(\equiv S(X))$ is the following set of 3! permutations:

$$\begin{pmatrix} 1 & 2 & 3 \\ 1 & 2 & 3 \end{pmatrix}, \begin{pmatrix} 1 & 2 & 3 \\ 2 & 3 & 1 \end{pmatrix}, \begin{pmatrix} 1 & 2 & 3 \\ 3 & 1 & 2 \end{pmatrix}, \begin{pmatrix} 1 & 2 & 3 \\ 1 & 3 & 2 \end{pmatrix}, \begin{pmatrix} 1 & 2 & 3 \\ 3 & 2 & 1 \end{pmatrix}, \begin{pmatrix} 1 & 2 & 3 \\ 2 & 1 & 3 \end{pmatrix}.$$

Now consider the equilateral triangle in Fig. D.1. This triangle is invariant under the six-element group of geometric transformations $C_{3v} = \{E, C_3, C_3^2, \sigma_v(1),$ $\sigma_v(2), \sigma_v(3)\}$ where E, C_3 and C_3^2 correspond respectively to counterclockwise rotations of the triangle by angles 0, $2\pi/3$ and $4\pi/3$ about the centroid 0, while $\sigma_v(1), \sigma_v(2)$ and $\sigma_v(3)$ correspond to reflections about 11', 22' and 33', respectively. By regarding the outcomes of these transformations as shufflings of the vertices 1, 2 and 3 of the triangle, we find that they are the same as the permutations of the

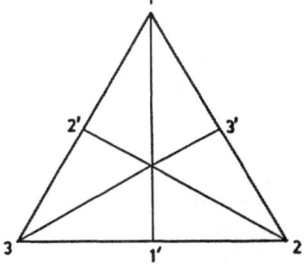

Fig. D.1. Illustration of C_{3v} as a transformation group acting on an equilateral triangle whose vertices are marked 1, 2 and 3. The lines 11', 22' and 33' are perpendiculars from the vertices on the opposite sides

three objects listed above. Thus C_{3v} is a (in this case, improper) subgroup of S_3 and is therefore a transformation group on $X = \{1, 2, 3\}$.

Let G be a group and X a set. Then G is said to *act on* X (on the left) if there is a mapping $G \times X \to X$ written $(g, x) \mapsto gx (g \in G, x \in X)$ such that

(i) $ex = x$ for all $x \in X$, where e is the identity element of G, and
(ii) $(gh)x = g(hx)$, where $g, h \in G$.

The set X is called the *group space* or *G-space*. Transformation groups can be thought of as special cases of group actions.

Consider the group action $G \times X \to X$. Two points $x, x' \in X$ are *equivalent* under G i.e. $x \sim x'^2$ if there exists ($\exists$) a $g \in G$ such that $gx = x'$. An *equivalence class* with the representative $x_0 \in X$ is the set

$$[x_0] = \{gx_0 | gx_0 \in X , \quad \text{for all} \quad g \in G\} .$$

Each such equivalence class is called an *orbit* of G. The set $[x_0]$ is the orbit of G generated by the element x_0. We give two examples.

Example 1: Let $X = \mathbb{R}^2$, the Euclidean plane, and let G be the group of all rotations about the origin. Under the action of G, the point $x \mapsto x'$ executes a circular arc subtending an angle θ corresponding to the amount of rotation. All points in the full circle around the origin and containing the point x, represent the orbit $[x]$.

Example 2: Consider the group C_{3v} acting on the set of vertices $\{1, 2, 3\}$ of the equilateral triangle (Fig. D.1). Under the action of C_{3v}, the vertex 1 is transformed into 2 and 3, i.e. the orbit of vertex 1 is the whole set $\{1, 2, 3\}$.

If for any pair of points $x, x' \in X$, $\exists$ at least one element of $g \in G$ such that $x' = gx$, then the action of G is said to be *transitive* and the set X is called a *homogeneous space*. Obviously if G acts on X transitively, then there is only one orbit in the whole of X.

If a point $x \in X$ is mapped into itself $(x \mapsto x)$ under all elements of G, then it is called a *fixed point*.

The subgroup of all elements of G leaving x fixed, $G_x \equiv \{g | g \in G, gx = x\}$, is called the *isotropy* (or *stability*, or *little*) group of x.

An inner automorphism of a group can be viewed as a group action. The former induces a map $G \times G \to G$ given by $(g, x) \mapsto gxg^{-1}$ for all $g, x \in G$. It is therefore an action of G on the set G. Conjugation is an equivalence relation on

---

[2] Note that the relation " $\sim$ " satisfies three properties: (i) $x \sim x$ (reflexive) (ii) $x \sim y$ implies $y \sim x$ (symmetric) and (iii) $x \sim y$, $y \sim z$ together imply $x \sim z$ (transitive). Any relation (defined on a set) satisfying these three properties is called an *equivalence relation*. A set with an equivalence relation can be partitioned into classes (called equivalence classes) of elements equivalent to one another.

the set G; i.e., $x \sim x'$ if $x' = gxg^{-1}$ for some $g \in G$. The quotient set $G/\sim$ consists of orbits $[x] = \{x'|x' = gxg^{-1}\}$ known as *conjugacy classes*.

Example 1: The group C_{3v} of the equilateral triangle has three conjugacy classes, $\{E\}$, $\{C_3, C_3^2\}$ and $\{\sigma_v(1), \sigma_v(2), \sigma_v(3)\}$. For example, $\sigma_v(1)$ and $\sigma_v(3)$ are conjugate to each other since

$$C_3 \sigma_v(1) C_3^2 = \sigma_v(3) \text{ , etc.}$$

Example 2: Consider the group of rotations and translations (the 2D Euclidean group E(2)) as a transformation group on $\mathbb{R}^2$. The rotations are given by $\varrho_\theta : x \to \underset{\sim}{R}(\theta)x$ and translations, by $\tau_a : x \to x + a$. Here $x \in \mathbb{R}^2$, $\underset{\sim}{R}(\theta)$ is the (orthogonal) rotation matrix corresponding to a rotation of angle θ and a is the vector of translation. Two translations commute: $t_b t_a t_b^{-1} = t_a$. Also,

$$(\varrho_\theta t_a \varrho_\theta^{-1})x = \varrho_\theta t_a (\underset{\sim}{R}(\theta)^{-1}x) = \varrho_\theta (\underset{\sim}{R}(\theta)^{-1}x + a)$$

$$= \underset{\sim}{R}(\theta)\underset{\sim}{R}(\theta)^{-1}x + \underset{\sim}{R}(\theta)a$$

$$= x + \underset{\sim}{R}(\theta)a = t_{\underset{\sim}{R}(\theta)a}x$$

for any $x \in \mathbb{R}^2$. This means $\varrho_\theta t_a \varrho_\theta^{-1} = t_{\underset{\sim}{R}(\theta)a}$. Therefore the set of translations form a single conjugacy class. However, given a rotation ϱ_θ, its conjugacy class consists of all transformations $t_b \varrho_\theta$ with arbitrary b. The reader would have noted that rotations act as automorphisms of translations in this example.

Let S be a subgroup of G. The group multiplication law defines a group action $G \otimes S \to G$ given by $(g, s) \mapsto gs$, called left multiplication by G. The orbits of S under this action are called *left cosets* of S in G. If, instead of the action $G \otimes S \to G$, we use $S \otimes G \to G$, we get the right cosets. Two elements $a, b \in G$ belong to the same left coset of S if and only if $b = as$ for some $s \in S$, i.e., if and only if $a^{-1}b \in S$. This is an equivalence relation.

Consider, as an example, the subgroup $S = \{E, \sigma_v(1)\}$ of the group C_{3v}. We note that $C_3 \sim \sigma_v(2)$ since $C_3 \sigma_v(1) = \sigma_v(2)$. Similarly, $C_3^2 \sim \sigma_v(3)$ since $C_3^2 \sigma_v(1) = \sigma_v(3)$. Therefore the left cosets of S in C_{3v} are $\{E, \sigma_v(1)\}$, $\{C_3, \sigma_v(2)\}$ and $\{C_3^2, \sigma_v(3)\}$.

A subgroup N of G is called a *normal* subgroup (denoted $N \lhd G$) if its left cosets coincide with its right cosets. It is also sometimes called an *invariant* subgroup. For the subgroup $S = \{E, \sigma_v(1)\}$ of C_{3v}, the left cosets are $\{E, \sigma_v(1)\}$, $\{C_3, \sigma_v(3)\}$ and $\{C_3^2, \sigma_v(2)\}$, which are different from the right cosets. Thus $\{E, \sigma_v(1)\}$ is not a normal subgroup. However, the left and right cosets of the subgroup $N = \{E, C_3, C_3^2\}$ coincide; they are $\{E, C_3, C_3^2\}$ $(= N)$, and $\{\sigma_v(1), \sigma_v(2), \sigma_v(3)\}$. Thus $N \lhd C_{3v}$.

Clearly, if $N \lhd G$, the quotient set $G/\sim$ has a group structure. This is the *factor group* or *quotient group*, written G/N. In the example just considered, $C_{3v}/N = \mathbb{Z}_2$. A group having no proper invariant subgroup (i.e., an invariant subgroup other than the group itself or the trivial group consisting of the identity element alone) is said to be *simple*. A group with no (proper) abelian invariant subgroup is *semisimple*.

We shall deal with continuous groups in the next topic (Sect. E.4).

E. A Brief Introduction to Homotopy and Lie Groups

The topological approach to the study of defects in ordered media requires the tools of homotopy theory and the theory of Lie groups. This Appendix offers an introduction to these topics. For further details, we recommend references [E.1–6].

E.1 Topology

In a sense, topology is a generalization of Euclidean geometry. According to Klein, a geometry $\Gamma(S, G)$ (and that includes Euclidean geometry) is given by specifying the underlying set S and the group G of transformations $G:S \rightarrow S$ acting on the set.

In the case of Euclidean geometry Γ^e, S is the set of all points (x, y) on the plane $\mathbb{R}^2$. The points can be represented by vectors $r = (x, y)$, $r' = (x', y')$ etc. which form a vector space. S is thus a metric space, i.e., a vector space with a metric

$$d(r, r') \equiv |r - r'| = \{(x - x')^2 + (y - y')^2\}^{1/2} .$$

The group G consists of:

Translations:	$r \mapsto r + a$,	$a = (a_1, a_2)$
Rotations:	$\begin{pmatrix} x \\ y \end{pmatrix} \mapsto \begin{pmatrix} \cos\theta & \sin\theta \\ -\sin\theta & \cos\theta \end{pmatrix} \begin{pmatrix} x \\ y \end{pmatrix}$	(E.1)
Reflections:	$(x, y) \mapsto (x, -y)$	
Dilations:	$(x, y) \mapsto c(x, y) \equiv (cx, cy).$	

In the above, a_1, a_2, θ and c are real parameters. The first three operations, called *isometries* or *rigid motions*, preserve distance between any two points. The fourth set of operations, called *homothety*, does not preserve the metric, although it preserves the angles between two lines.

The group G, referred to as the *principal group* of the geometry, is, in this case, the group of transformations consisting of all isometries and homotheties. This group is also called the *group of similarities*. Euclidean geometry is the study of properties which are invariant under the similarity group.

The choice of a particular set S and the corresponding group G may be called a model of the geometry $\Gamma(S, G)$.

We now come to the concept of a neighbourhood. In the Euclidean plane, surrounding any point $r = (x, y)$, there is a set $N(r, \delta) = \{r' | r' \in \mathbb{R}^2, d(r, r') < \delta\}$ of points, which define the interior of a disc-shaped region of radius δ ($\delta > 0$, $\delta \in \mathbb{R}$). This set is called an *open set* or a *neighbourhood* around r. The set $B(r, \delta) = \{r' | r' \in \mathbb{R}^2, d(r, r') = \delta\}$ consists of points on the boundary of $N(r, \delta)$. Given any two points r and r' ($\neq r$) however close, it is always possible to choose two open sets surrounding r and r' which do not overlap. This intuitively obvious fact is generalized in topology.

In topology, we have to deal with the notion of continuous deformation, and this is where the concept of neighbourhood becomes important. In a metric space, the concept of distance is used for defining a neighbourhood. However, strange as it may sound, the concept of distance is not necessary for defining nearness and continuity. What is more crucial is the concept of an open set. This leads us to the following definition:

Let X be a non-empty set. A class $\mathcal{T}$ of subsets of X is a *topology* on X if and only if $\mathcal{T}$ satisfies the conditions:

i) X and the null set $\varnothing$ belong to $\mathcal{T}$,
ii) the union of any number of sets in $\mathcal{T}$ belongs to $\mathcal{T}$, and
iii) the intersection of any two sets in $\mathcal{T}$ belongs to $\mathcal{T}$.

The members of $\mathcal{T}$ are called *open sets*, and the pair $(X, \mathcal{T})$ is called a *topological space*. The simplest example of a topological space is the familiar real line. Let $X = \mathbb{R}$, the real line, and let $\mathcal{U}$ be the class of all open intervals in $\mathbb{R}$ of the form $(a, b) \equiv \{x | a < x < b\}$. The pair $(\mathbb{R}, \mathcal{U})$ is a topological space. As another (abstract) example, consider the set $X = \{a, b, c, d, e\}$, and let

$$\mathcal{T} = \{X, \varnothing, \{a\}, \{c, d\}, \{a, c, d\}, \{b, c, d, e\}\} \ .$$

It is easy to see that all the three definitions of a topological space are satisfied.

The *neighbourhood* of a point $x \in X$ in a topological space is the set $N(x)$ containing an open set U (i.e. $N(x) \supset U$) such that $x \in U \subset N(x)$. The concept of continuity is made precise in the following way. Let $(X, \mathcal{T})$ and $(Y, \mathcal{U})$ be two topological spaces. A function $f: X \to Y$ is continuous relative to $\mathcal{T}$ and $\mathcal{U}$ if and only if $S \in \mathcal{U}$ implies

$$f^{-1}(S) = \{x | x \in X, f(x) \in S\} \in \mathcal{T} \ .$$

Analogous to the concept of morphisms for groups, we have here the important concept of *homeomorphism*. Two topological spaces $(X, \mathcal{T})$ and $(Y, \mathcal{U})$ are said to be *homeomorphic* or *topologically equivalent* if there exists a continuous function $f: X \to Y$ which is a bijection (i.e., one-to-one and onto). The function f is called a *homeomorphism*.

As an example, consider $(\mathbb{R}, \mathcal{U})$, the real line with the usual topology, and let $S = \{x | x \in \mathbb{R}, -1 < x < 1\}$ be an open interval. Let $\mathcal{U}'$ be the class of all open sets of S. (Clearly $\mathcal{U}'$ is obtained from $\mathcal{U}$ by taking the intersection of S with each member of $\mathcal{U}$.) Then the topological spaces $(\mathbb{R}, \mathcal{U})$ and $(S, \mathcal{U}')$ are homeomorphic to each other. A homeomorphism is established by the function $f: S \to \mathbb{R}$ given by $x \mapsto f(x) = \tan(\pi x/2)$. Topologically, $\mathbb{R}$ and S are equivalent.

The set of homeomorphisms forms a group. Indeed the group of homeomorphisms is the principal group of topology. All theorems of topology, once proved for a given topological space, remain true under homeomorphism. With additional structures, a topological space can be turned into a continuous group (see Sect. E.3).

E.2 Elements of Homotopy Theory

Homotopy is a branch of algebraic topology. The main problem in topology can be stated as follows: Given two topological spaces X, Y, how does one know if they are homeomorphic to each other? The answer to this seemingly simple question is really very difficult. Mathematicians have converted this problem into a simpler one. The trick is to associate various group structures denoted by $\Pi_n(X)$ with a given space X. The group $\Pi_n(X)$ is known as the nth *homotopy group* of X $(n = 1, 2, 3, \ldots)$; the first homotopy group is also called the *fundamental group* of X. If $\Pi_n(X)$ is not isomorphic to $\Pi_n(Y)$ for *all* n, then X and Y cannot be homeomorphic. However, the converse is not necessarily true. We now introduce some necessary concepts essential to proceed further.

A *path* $\sigma(t)$ from a point x_1 to a point x_2 in a topological space X is a continuous map of the unit interval $0 < t < 1$ (written $[0, 1] \equiv I$, for brevity) into X, $\sigma: I \to X$, such that $\sigma(0) = x_1$ and $\sigma(1) = x_2$.

Let $\sigma_0(t)$ and $\sigma_1(t)$ be two paths connecting x_1 and x_2. Then σ_0 is said to be *homotopic* to σ_1, written $\sigma_0 \simeq \sigma_1$, if there exists a continuous function $h: I \to X$ parametrized by a real variable $t \in I$ such that

$$h_t(0) = \sigma_0(t) , \qquad h_t(1) = \sigma_1(t) , \quad \text{and} \tag{E.2}$$

$$h_0(s) = x_1 , \qquad h_1(s) = x_2 ,$$

for all $t \in I$ and all $s \in I$. This is indicated in Fig. E.1. $h_t(s)$ is called the *homotopy* connecting the paths $\sigma_0(t)$ and $\sigma_1(t)$. Alternatively, one says that $\sigma_0(t)$ can be continuously deformed into $\sigma_1(t)$.

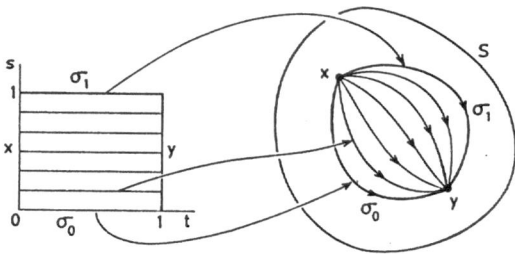

Fig. E.1. Illustration of homotopy mapping σ_0 to σ_1 continuously

A space X is *arcwise connected*, or *pathwise connected* if every pair of points of X can be connected by a path in X.

A pathwise connected space X is *simply connected* if every pair of paths σ_0, σ_1 in X, connecting any two arbitrary points x_1 and x_2 of X, are homotopic.

The relation "two-points x_1 and x_2 can be connected by a path in X" is an *equivalence relation* ($\simeq$). Hence paths in X can be decomposed into equivalence classes.

Members of equivalence classes in X, generated by the relation $\simeq$ (i.e. $X/\simeq$), are called *path components* of X. Thus a simply-connected space consists of a single path component.

A special type of path playing an important role in homotopy theory is a loop. A *loop* in a space X is a path $\sigma: I \to X$ such that $\sigma(0) = \sigma(1)$. The point $\sigma(0) = \sigma(1) = x$ is called the *base point* of the loop. The space $\Omega(X)$ of all loops in X contains a subspace $\Omega(X, x)$ of x-based loops in X.

E.2.1 The First Homotopy Group

The space $\Omega(X, x)$ can be given a group structure which is of paramount importance to us. The underlying set of this group is the space $\Omega(X, x)/\simeq$, i.e. the distinct equivalence classes of loops. Let $[\alpha(t)], [\beta(t)] \in \Omega(X, x)/\simeq$. Thus $[\alpha(t)]$ is the set of all loops based at x and homotopically equivalent to the representative loop $\alpha(t)$, etc. Then a group composition law can be defined between $[\alpha]$ and $[\beta]$ to produce another element

$$[\gamma] = [\alpha] \circ [\beta] \in \Omega(X, x)/\simeq .$$ (E.3)

In order to define $[\alpha] \circ [\beta]$, we first introduce multiplication of the two loops α and β. The product γ of two x-based loops α and β is written as $\gamma = \alpha * \beta$ and is defined by

$$\gamma(t) = \begin{cases} \alpha(2t) , & 0 \leqslant t \leqslant 1/2 \\ \beta(2t-1) , & 1/2 \leqslant t \leqslant 1. \end{cases}$$ (E.4)

Intuitively, $\alpha * \beta$ is the loop at x traced by moving successively along the loops α and β (Fig. E.2). The above definition induces naturally the composition law $[\alpha] \circ [\beta]$ in the space $\Omega(X, x)/\simeq$. On the right in Fig. E.2 is shown another loop γ' which is homotopically equivalent to γ. This loop is obtained by detaching from x the middle portion of γ corresponding to the end of α and the beginning of β, making sure that the resulting loop starts and terminates at x. Note that

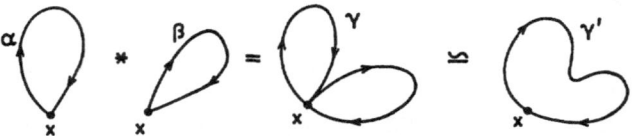

Fig. E.2. The product of loops α and β at x

removing the point x from γ results in two separate pieces, whereas removal of x from γ' results in a single piece. Thus, γ and γ', though homotopic to each other, are *not* homeomorphic (since homeomorphism should preserve connectivity).

The product law $[\alpha] \circ [\beta]$ induced by the multiplication $\alpha * \beta$ is given by

$$[\alpha] \circ [\beta] = [\alpha * \beta] \ . \tag{E.5}$$

We now state (without proof) an important theorem [E.7]: The set $\Omega(X, x)/\simeq$ with the composition law defined by (E.5) is a group denoted $\Pi_1(X, x)$. In this group, the identity element $[\delta]$ is defined as the set

$$[\delta] = \{\alpha | \alpha \simeq \delta \ , \quad \delta(t) = x \text{ for all } t \in I\} \ , \tag{E.6}$$

i.e., the set of all degenerate loops. The meaning of the inverse of an element is understood with the help of the relation $[\alpha]^{-1} = [\alpha^{-1}]$, where the loop $\alpha^{-1}(t)$ is obtained from the loop $\alpha(t)$ by reversing the direction of traverse;

$$\alpha^{-1}(t) \equiv \alpha(1 - t) \tag{E.6}$$

(Fig. E.3).

There is an important result elucidating the dependence of the group $\Pi_1(X, x)$ on the base point x. If X is path-connected, then $\Pi_1(X, x_1)$ is isomorphic to $\Pi_1(X, x_2)$ for any two different points $x_1, x_2 \in X$, i.e., in a path-connected space, there is a unique $\Pi_1(X)$. The isomorphism between $\Pi_1(X, x_1)$ and $\Pi_1(X, x_2)$ is established via a path σ connecting x_1 and x_2. Given a loop α at x_1, one can associate a loop $\sigma_\#(\alpha) = \sigma\alpha\sigma^{-1}$, based at x_2, defined by

$$(\sigma\alpha\sigma^{-1})(t) = \begin{cases} \sigma(3t) \ , & 0 \leqslant t \leqslant 1/3 \\ \sigma(3t - 1) \ , & 1/3 \leqslant t \leqslant 2/3 \\ \sigma(3t - 2) \ , & 2/3 \leqslant t \leqslant 1 \ . \end{cases} \tag{E.7}$$

This is illustrated in Fig. E.4. This means that the loop spaces $\Omega(X, x_1)$ and $\Omega(X, x_2)$ are related by the isomorphism map $\sigma_\#$. Because of this, there arises an isomorphism $\sigma_\#: \Pi_1(X, x_1) \to \Pi_1(X, x_2)$. In this sense, the first homotopy group is a topological invariant of the path-connected space X. The independence of $\Pi_1(X, x)$ on x is often referred to as *free homotopy* [E.1].

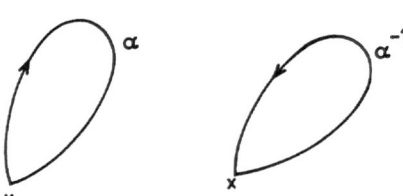

Fig. E.3. The loop α^{-1} is the inverse of the loop α

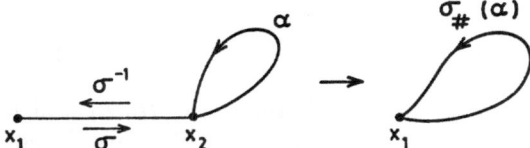

Fig. E.4. Illustration of the transfer of a loop based at x_2 to another point x_1 by using the map σ_*

E.2.2 Higher Homotopy Groups

The groups $\Pi_n(X, x)$ for $n > 1$ are obtained by the generalization of the first homotopy group. Thus, $\Pi_n(X, x)$ consists of elements which are the equivalence classes of n-dimensional loops possible in X at the point x. A precise formulation can be achieved in the following manner. We note that a loop α is a map $\alpha : I \to X$ such that $\alpha(\partial I) = x$, where ∂I, the boundary of I, is the set $\{0, 1\}$. The concept of a loop can now be generalized. Let I^n denote the n-fold Cartesian product of I, i.e. the n-cube. Let $(t_1, t_2, \ldots, t_n) \in I^n$. An $(n-1)$-face of I^n is obtained by setting some coordinate t^i to be 0 or 1. The boundary ∂I^n of I^n is the union of all the $(n-1)$-faces. Clearly, ∂I^n is homeomorphic to S^{n-1}. (The reader is urged to verify all these statements for $n = 2$ and 3). Let $\Omega^n(X, x)$ be the set of all maps $\alpha : I^n \to X$ which also maps $\partial I^n \to x$, i.e.,

$$\Omega^n(X, x) = \{\alpha \,|\, \alpha : I^n \to X \quad \text{and} \quad \alpha : \partial I^n \to x, x \in X\} \ . \tag{E.8}$$

Homotopic equivalence ($\simeq$) of loops in $\Omega^n(X, x)$ is defined analogously to the previous case of $n = 1$; $[\alpha]$, now, is the class of "loops" containing the representative loop α. A loop product is now defined, in the space $\Omega^n(X, x)$, analogous to (E.5).

$$\gamma(t) = (\alpha * \beta)(t) = \begin{cases} \alpha(2t_1, t_2, \ldots, t_n) \ , & 0 \leq t_1 \leq 1/2 \\ \beta(2t_1 - 1, t_2, \ldots, t_n) \ , & 1/2 \leq t_1 \leq 1 \end{cases} \tag{E.9}$$

for all $t = (t_1, t_2, \ldots, t_n) \in I^n$. Generalization of the loop product in the space $\Omega^n(X, x)/\simeq$ proceeds along lines similar to (E.5), with suitable reinterpretation of the symbols. Let $[\delta]$ denote the degenerate class of constant or trivial loops $\delta : I^n \to x$, and let the inverse $[\alpha]^{-1}$ of $[\alpha]$ be defined as $[\alpha]^{-1} = [\alpha^{-1}]$, where $\alpha^{-1}(t_1, t_2, \ldots, t_n) = \alpha(1 - t_1, t_2, \ldots, t_n)$. The group $\Pi_n(X, x)$ can now be defined as the space $\Omega^n(X, x)/\simeq$ of path-components with the composition law $[\alpha] \circ [\beta]$. The concept of free homotopy is defined by a trivial generalization of the previous case.

The following results [E.1] are very important.

Theorem:

The groups $\Pi_n(X, x)$ are all Abelian for $n > 1$. $\tag{E.10}$

Theorem:

$$\Pi_n(X \times Y, x) = \Pi_n(X, x) \otimes \Pi_n(Y, x) \ . \tag{E.11}$$

For the sake of completeness, we introduce also the object $\Pi_0(X)$, which stands for the set of path components (disjoint components) of X. We do not discuss the various methods for computing homotopy groups of a given space (see references [E.4, 5, 7]). However, Table E.1 gives a list of the homotopy groups $\Pi_n(x)$ $(n = 1, 2, 3)$ for a few commonly occurring spaces. We next introduce the concepts of continuous groups (or topological groups) and of Lie groups in particular.

Table E.1. The homotopy groups $\Pi_n(X)$ $(n = 1, 2, 3)$ for a few commonly occurring spaces. The symbols are $-\{e\}$: the trivial group consisting only of the identity; $\mathbb{Z}$: the group of the integers; Q: the quaternion group; D_n: the dihedral group of order n; and $\approx$ denotes isomorphism

X	$\Pi_1(X)$	$\Pi_2(X)$	$\Pi_3(X)$
S^1	$\mathbb{Z}$	$\{e\}$	$\{e\}$
S^2	$\{e\}$	$\mathbb{Z}$	$\mathbb{Z}$
$S^3 \approx SU(2)$	$\{e\}$	$\{e\}$	$\mathbb{Z}$
$\mathbb{R}P(3) \approx SO(3)$	$\mathbb{Z}_2$	$\{e\}$	$\mathbb{Z}$
$\mathbb{R}P(2) \approx SO(3)/D_\infty$	$\mathbb{Z}_2$	$\mathbb{Z}$	$\mathbb{Z}$
$SO(3)/D_2$	Q	$\{e\}$	$\mathbb{Z}$

E.3 Continuous Groups and Lie Groups

Consider a topological space G which is locally Euclidean, i.e., every point $a \in G$ can be assigned coordinates $a^1, \ldots, a^r$ of E^r. Such a space is called a *manifold* of dimension r. (The surface of a sphere is a "smooth" 2D manifold.) Let the elements of G have group properties too. Then if $a, b \in G$, their group product $c = ab$ should be "continuous" (and so also should be the inverse a^{-1} of a). More precisely, we can write the coordinates of c:

$$c^\alpha = \varphi^\alpha(a^1, \ldots, a^r, b^1, \ldots, b^r) = \varphi^\alpha(a, b) \ , \tag{E.12}$$

where the function φ^α $(\alpha = 1, \ldots, r)$ is continuous in its arguments. The group properties may be stated in terms of the functions φ^α:

1) Closure: $c = ab$; meaning

$$c^\alpha = \varphi^\alpha(a, b), c \in G \ , \tag{E.13a}$$

2) Associativity: $a(bc) = (ab)c$; meaning

$$\varphi^\alpha[a, \varphi(b, c)] = \varphi^\alpha[\varphi(a, b), c] \ , \tag{E.13b}$$

3) Identity: $ea = a = ae$; meaning

$$\varphi^{\alpha}(e, a) = a^{\alpha} = \varphi^{\alpha}(a, e) \ , \tag{E.13c}$$

4) Inverse: $aa^{-1} = e = a^{-1}a$; meaning

$$\varphi^{\alpha}(a, a^{-1}) = e^{\alpha} = \varphi^{\alpha}(a^{-1}, a) \ . \tag{E.13d}$$

It is the *continuous group of transformations* (CGT), rather than the abstract continuous group, that is of importance for applications. A CGT is nothing but an r-parameter (or an r-dimensional) continuous group G acting on an n-dimensional manifold M (we may refer to it as a "geometric space"). The group action $f: G \times M \to M$ (a left action) is precisely in the sense introduced in Appendix D, except that f is a continuous function of its arguments. Let a typical point $x \in M$ have coordinates $\{x^i | i = 1, \ldots, n\}$. The group multiplication function $\varphi: G \times G \to G$ obeys the properties (E.13) and the functions f^i have the following properties: (in a condensed notation, the functions f and φ can be dropped)

1) Closure: $a \in G$, $x \in M$ implies $ax \in M$, meaning

$$y^i = f^i(a^1, \ldots, a^r; x^1, \ldots, x^n) \in M \ . \tag{E.14a}$$

2) Associativity: $a(bx) = (ab)x$, meaning

$$f^i[a, f(b, x)] = f^i[\varphi(a, b), x] \ . \tag{E.14b}$$

3) Identity: $ex = x$, meaning

$$f^i(e, x) = x^i \ . \tag{E.14c}$$

4) Inverse: $a^{-1}(ax) = a(a^{-1}x) = (aa^{-1})x = x$, meaning

$$f^i[a^{-1}, f(a, x)] = f^i[a, f(a^{-1}, x)] = f^i[\varphi(a, a^{-1}), x] = x^i \ . \tag{E.14d}$$

The right action $M \times G \to M$ is defined similarly. If for each x we have the isotropy group $G_x = \{e\}$, then the group action is called *free* and if $ax = x$ implies $a = e$, then the action is called *effective*. All concepts such as transitive action, homogeneous G-space, conjugacy, automorphism, normal subgroup, coset decomposition, etc. introduced in Appendix D, retain their meaning in the present context.

Nothing was mentioned regarding the connectivity properties of the parameter space of G. In general, it may consist of several path components, described by the group $\Pi_0(G)$. Let G_0 be the path component of G connected to the identity element $e \in G$. The following facts, stated as theorems, are extremely important in the study of continuous groups.

Theorem:

G_0 is an invariant subgroup of G.   (E.15)

Theorem:

The quotient group (or the factor group) G/G_0 is a discrete subgroup D of dimension zero and $\Pi_1(G) = D$.   (E.16)

Because of theorem (E.16), the structure of the entire continuous group G is known if we only know the structure of the group G_0 and the discrete group $\Pi_1(G)$.

A *Lie group* is the connected component of a continuous group. Thus the group G_0 is a Lie group. Quite a lot of information regarding a Lie group G is contained in the neighbourhood of its identity e. G is called a *local Lie group of dimension r* if a parametrization (i.e., a coordinate system) can be made such that (1) $a^\alpha = 0$ for $a = e$ ($\alpha = 1, \ldots, r$), and, (2) in some neighbourhood of e, the coordinates $\{c^\alpha\}$ of the product $c = ab$ ($a, b \in G$) are continuous functions of $\{a^\alpha\}$ and $\{b^\alpha\}$. Let $c^\alpha = \varphi^\alpha(a, b)$, as in (E.13). A *Lie group of transformations* acting on an (analytic) manifold M is described by (analytic) functions $\{f^i\}$ satisfying (E.14). Let us consider the Lie group action given by $f: G \times M \to M$:

$$y^i = f^i(a^1, \ldots, a^r; x^1, \ldots, x^n) \equiv f^i(a, x) \ . \tag{E.17}$$

When $a = e$, (E.17) reduces to $x = f(0, x)$. Consider the evolution of the point $x = f(0, x)$ under a small change δa of the parameter a. Owing to the continuity of f, the point $x \in M$ changes by a small amount dx, $x \mapsto x + dx$, where (using a Taylor expansion to first order in δa)

$$x^i + dx^i = f^i(\delta a, x) = f^i(0, x) + \left.\frac{\partial f^i(b, x)}{\partial b^\alpha}\right|_{b=0} \delta a^\alpha \tag{E.18}$$

The continuity of the group composition function φ implies that the point $a = \varphi(0, a)$ changes to a nearby point $(a + da) \in G$:

$$a^\alpha + da^\alpha = \varphi^\alpha(\delta a, a) = \varphi^\alpha(0, a) + \left.\frac{\partial \varphi^\alpha(b, a)}{\partial b^\beta}\right|_{b=0} \delta a^\beta \ , \tag{E.19}$$

again using a Taylor expansion. Since $f(0, x) = x$ and $\varphi(0, a) = a$, we obtain

$$da^\alpha = \mu_\beta^\alpha(a)\,\delta a^\beta \ , \quad \mu_\beta^\alpha(a) = \left.\frac{\partial \varphi^\alpha(b, a)}{\partial b^\beta}\right|_{b=0} \tag{E.20}$$

and

$$dx^i = u_\alpha^i(x)\,\delta a^\alpha \ , \quad u_\alpha^i(x) = \left.\frac{\partial f^i(b, x)}{\partial b^\alpha}\right|_{b=0} \ . \tag{E.21}$$

If we now consider arbitrary functions $\psi(a)$ and $F(x)$ in the spaces G and M respectively, then the effect of small changes $a \mapsto a + \delta a$ and $x \mapsto x + dx$ is to cause corresponding changes in ψ and $F: \psi \to \psi + d\psi$ and $F \to F + dF$, where

$$d\psi(a) = \frac{\partial \psi}{\partial a^{\alpha}} da^{\alpha} = \mu_{\beta}^{\alpha}(a) \delta a^{\beta} \frac{\partial \psi}{\partial a^{\alpha}} = \delta a^{\beta} X_{\beta} \psi(a) \ , \tag{E.22}$$

and

$$dF(x) = dx^{i} \frac{\partial F}{\partial x^{i}} = \delta a^{\alpha} u_{\alpha}^{i}(x) \frac{\partial F}{\partial x^{i}} = \delta a^{\alpha} \tilde{X}_{\alpha} F(x) \ . \tag{E.23}$$

The operators X_{α} and $\tilde{X}_{\alpha}$ are called the *infinitesimal generators of the group* and the *infinitesimal generators of transformation*, respectively.

For any two elements a and b in the neighbourhood of the identity, a closer look at the composition function reveals that it must have the form

$$\varphi^{\alpha}(a, b) = a^{\alpha} + b^{\alpha} + f_{\mu\nu}^{\alpha} a^{\mu} b^{\nu} + \ldots \ . \tag{E.24}$$

We now state an important theorem which follows from the associativity of φ. The antisymmetric part of $f_{\mu\nu}^{\alpha}$,

$$C_{\mu\nu}^{\alpha} = -C_{\nu\mu}^{\alpha} = f_{\mu\nu}^{\alpha} - f_{\nu\mu}^{\alpha} \tag{E.25}$$

obeys the Jacobi identity

$$C_{\alpha\beta}^{\sigma} C_{\sigma\gamma}^{\varrho} + C_{\beta\gamma}^{\sigma} C_{\sigma\alpha}^{\varrho} + C_{\gamma\alpha}^{\sigma} C_{\sigma\beta}^{\varrho} = 0 \ . \tag{E.26}$$

$\{C_{\mu\nu}^{\alpha}\}$ are called *structure constants* of the group. They completely determine the structure of G, i.e., given the $\{C_{\mu\nu}^{\alpha}\}$, the expansion of $\varphi(a, b)$ can be found to *any* order in a and b. It turns out that the infinitesimal generators of transformation $\{\tilde{X}_{\alpha}\}$ obey the relations

$$[\tilde{X}_{\alpha}, \tilde{X}_{\beta}] = \tilde{X}_{\alpha} \tilde{X}_{\beta} - \tilde{X}_{\beta} \tilde{X}_{\alpha} = -[\tilde{X}_{\beta}, \tilde{X}_{\alpha}] = C_{\alpha\beta}^{\gamma} \tilde{X}_{\gamma} \ , \tag{E.27}$$

and

$$[[\tilde{X}_{\alpha}, \tilde{X}_{\beta}], \tilde{X}_{\gamma}] + \text{cyclic permutations} \equiv 0 \ . \tag{E.28}$$

Identical relations hold good for X_{α}. The relation (E.28) is yet another form of the Jacobi identity (E.26). The linear space spanned by $\{\tilde{X}_{\alpha}\}$ with the commutator product $[.,.]$ is an algebra called the *Lie algebra* $\mathcal{G}$ of the Lie group of transformations G. The generators of $\mathcal{G}$ are used in the construction of internal covariant derivatives which are of fundamental importance in gauge theories (see Appendix F). We recall that in the Poisson bracket method of deriving hydrodynamic equations (Sect. A.4), the Poisson bracket $\{l_{\alpha}, l_{\beta}\}$ of the generators $\{l_{\alpha}\}$ is a Lie bracket satisfying (E.27, 28).

We now consider an example, where our concern is to obtain a matrix representation of the Lie algebra of SO(3). The elements of the group SO(3) are formed by 3×3 orthogonal matrices A of determinant $+1$, where I is the identity matrix. These matrices act on the geometric space $M = E^3$. An infinitesimal rotation is a transformation of the form

$$\underset{\sim}{A} = \underset{\sim}{I} + \underset{\sim}{B} ,$$
(E.29)

where B is a matrix having all its elements infinitesimally close to zero. The requirement $A^T A = I$ implies that $B^T = -B$ and also that B is a first order infinitesimal quantity:

$$\underset{\sim}{B} = \begin{pmatrix} 0 & \delta\omega^3 & -\delta\omega^2 \\ -\delta\omega^3 & 0 & \delta\omega^1 \\ \delta\omega^2 & -\delta\omega^1 & 0 \end{pmatrix} .$$
(E.30)

If we write x as a column vector $(x^1, x^2, x^3)^T$, then we may write

$$\begin{pmatrix} dx^1 \\ dx^2 \\ dx^3 \end{pmatrix} = \begin{pmatrix} 0 & \delta\omega^3 & -\delta\omega^2 \\ -\delta\omega^3 & 0 & \delta\omega^1 \\ \delta\omega^2 & -\delta\omega^1 & 0 \end{pmatrix} \begin{pmatrix} x^1 \\ x^2 \\ x^3 \end{pmatrix} .$$
(E.31)

Identifying ω^α with the group parameters $\{a^\alpha\}$ and using (E.18), we get $dx^i = \varepsilon_{ij\alpha} x^j \delta a^\alpha (dx^1 = x^2 \delta a^3 - x^3 \delta a^2$, etc.). Since the generators $\tilde{X}_\alpha = u^i_\alpha \partial/\partial x^i$ (where u^i_α is defined by the relation $dx^i = u^i_\alpha \delta a^\alpha$), we get $\tilde{X}_\alpha = \varepsilon_{\alpha jk} x^j \partial/\partial x^k$. These $\{\tilde{X}_\alpha\}$ may be identified with the usual angular momentum operators. It is clear from (E.30) that the matrix representation for $\tilde{X}_\alpha$ is provided by

$$\tilde{X}_1 = \begin{pmatrix} 0 & 0 & 0 \\ 0 & 0 & 1 \\ 0 & -1 & 0 \end{pmatrix} , \quad \tilde{X}_2 = \begin{pmatrix} 0 & 0 & -1 \\ 0 & 0 & 0 \\ -1 & 0 & 0 \end{pmatrix} ,$$

$$\tilde{X}_3 = \begin{pmatrix} 0 & 1 & 0 \\ -1 & 0 & 0 \\ 0 & 0 & 0 \end{pmatrix} .$$
(E.32)

These generators of transformations are closed under the Lie algebra with the commutation relations

$$[\tilde{X}_\alpha, \tilde{X}_\beta] = \varepsilon_{\alpha\beta\gamma} \tilde{X}_\gamma .$$
(E.33)

The generators $\tilde{X}_\alpha$ of the Lie algebra of a Lie group describe the action, on the geometric space, of a group element infinitesimally close to the identity. Since the Lie group is a continuous group, the process can be repeated *ad infinitum* to generate the action of *any* group element. It is a theorem that any group element can be expressed as the matrix $\exp(a^\alpha \tilde{X}_\alpha)$ which is defined by the formal power series expansion of the exponential function.

There may exist several Lie groups $\{G_i\}$ sharing the *same* Lie algebra $\mathcal{G}$. For example, the group SU(2) of all unimodular 2×2 matrices with complex entries, has the same Lie algebra (E.33) as SO(3). (The generators of SU(2) are $i\sigma_1/2$, $i\sigma_2/2$ and $i\sigma_3/2$, where the $\{\sigma_\alpha\}$ are the familiar Pauli matrices.) Thus, sharing the same Lie algebra *does not* imply that the groups are isomorphic. The parameter spaces may be different for different G_i.

We now consider the parameter space of SO(3). The matrix $\underset{\sim}{A}(\omega)$ corresponding to a nontrivial set of parameters $a^\alpha = \omega^\alpha$ can be written as

$$\underset{\sim}{A}(\omega) = \exp(\omega^\alpha \tilde{X}_\alpha) \ . \tag{E.34a}$$

Using the generators $\tilde{X}_\alpha$ given by (E.32), we get

$$\underset{\sim}{A}(\omega) \equiv \underset{\sim}{A}(\hat{\omega}, \omega) = \underset{\sim}{I}_3 \cos \omega + \underset{\sim}{\Omega}(1 - \cos \omega) + (\hat{\omega}^\alpha \tilde{X}_\alpha) \sin \omega \tag{E.34b}$$

where

$$\omega = [(\omega^1)^2 + (\omega^2)^2 + (\omega^3)^2]^{1/2} \ , \quad \hat{\omega}^i = \omega^i/\omega \ , \quad \hat{\omega} = \omega/\omega \ ,$$

and

$$\Omega_{ij} = \delta_{i\alpha} \delta_{j\beta} \hat{\omega}^\alpha \hat{\omega}^\beta \ . \tag{E.34c}$$

This matrix describes the change of an orthonormal basis in E^3 brought about by a rotation through an angle ω about an axis in the $\hat{\omega}$ direction. It is evident from (E.34) that

$$\underset{\sim}{A}(\hat{\omega}, \omega + 2\pi) = \underset{\sim}{A}(\hat{\omega}, \omega) \ . \tag{E.34d}$$

Equation (E.34d) implies that the parameter space of SO(3) is a *solid sphere* of radius $\pi(0 \leqslant \omega \leqslant \pi)$, with the antipodal points identified $[\underset{\sim}{A}(\hat{\omega}, \pi) = \underset{\sim}{A}(-\hat{\omega}, \pi)]$. Such a space is denoted by $\mathbb{R}P(3)$, the real projective space of three dimensions.

It is evident that this space is pathwise connected. The detailed nature of the connectivity of SO(3) can be discussed with the help of Fig. E.5. Consider a loop L_1 which starts from the origin O of the sphere and closes *without* touching the surface of the sphere. It is clear that L_1 is homotopic to the origin. Now consider the curve L_2. This curve has two pieces from O to a point P on the surface of the sphere and from P' (the antipodal point of P) to O. In view of the identification of P and P', L_2 is clearly a loop. L_2 cannot be contracted to the origin and hence loops of type L_2 form a distinct class. It can be shown that all the loops possible in SO(3) are homotopic to either L_1 or L_2. Hence SO(3) is doubly connected, and $\Pi_1[SO(3)] = \mathbb{Z}_2$.

The connectivity of SU(2) can be discussed in a similar manner. An arbitrary element $\underset{\sim}{Q}(\omega)$ of SU(2) can be written as

$$\underset{\sim}{Q}(\omega) = \exp(\omega^\alpha \tilde{X}_\alpha) \ , \tag{E.35a}$$

where the ω^α are given by (E.34c) and the generators of transformation $\tilde{X}_\alpha$ are

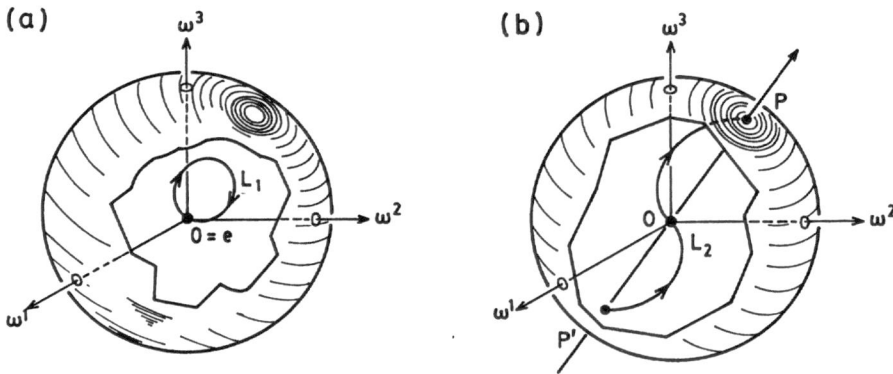

Fig. E.5. The parameter space of SO(3): (a) trivial loop L_1, (b) a loop L_2 which cannot be shrunk to the origin. Further explanations in the text

defined as

$$\widetilde{X}_\alpha = \tfrac{1}{2}i\underset{\sim}{\sigma}_\alpha \ . \tag{E.35b}$$

Using the properties of the Pauli matrices $\underset{\sim}{\sigma}_\alpha$, (E.35a) simplifies to

$$\underset{\sim}{Q}(\boldsymbol{\omega}) = \underset{\sim}{Q}(\hat{\boldsymbol{\omega}}, \omega) = \underset{\sim}{I}_2 \cos\frac{\omega}{2} + i(\hat{\boldsymbol{\omega}}^\alpha \underset{\sim}{\sigma}_\alpha)\sin\frac{\omega}{2} \ . \tag{E.35c}$$

From (E.35c) it is clear that

$$\underset{\sim}{Q}(\hat{\boldsymbol{\omega}}, \omega) = \underset{\sim}{Q}(\hat{\boldsymbol{\omega}}, \omega + 4\pi)$$
$$= -\underset{\sim}{Q}(\hat{\boldsymbol{\omega}}, \omega + 2\pi) \ . \tag{E.35d}$$

From this we infer that the parameter space of SU(2) is a solid sphere of radius 2π $(0 \leqslant \omega \leqslant 2\pi)$. It is important to note that *all* the elements $\underset{\sim}{Q}(\hat{\boldsymbol{\omega}}, 2\pi)$, on the surface of the sphere, correspond to one *single* group operation $-e$.

We now discuss the connectivity of SU(2). It is clear that all loops which do not touch the surface of the sphere are homotopic to the identity. Therefore we consider only the nontrivial loop shown in Fig. E.6. Since the whole of the surface of the sphere represents one single element, we may continuously deform the segment OCB of the loop so that it eventually falls upon the other segment OAB (through operations OB$' \to$ OB$'' \to$ OB$'''$ etc.). Thus the loop shown in the figure can be made homotopic to the loop OABAO which is homotopic to 0. Thus all loops in SU(2) can be shrunk to a point, and therefore it is *simply connected*. It is a theorem that among all the Lie groups having the same Lie algebra, there is always one which is simply connected (i.e., the Π_1 of this group is trivial). This group is called the *universal covering group*. In the above example, SU(2) is the

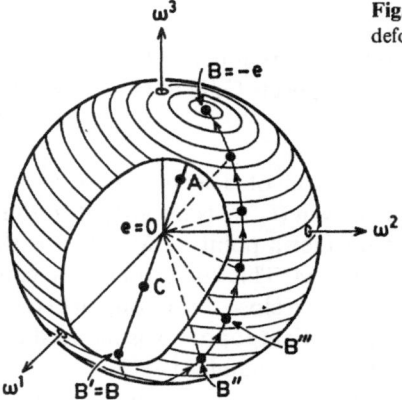

Fig. E.6. The parameter space of SU(2) showing the deformation of a loop to the origin

universal covering group of SO(3). It may be noted that SU(2) can also be parametrized in another manner, as done earlier in (4.3). From (4.4), the parameter space of SU(2) is the sphere S^3 (of radius unity). All loops in S^3 can be shown to be equivalent to a point.

The concept of a *discrete* Lie group action on a manifold is also important. By a discrete group D, we mean a group with a countable number of elements, where every element is an open set. (Thus D has a discrete topology). As a manifold, D is of dimension zero; thus D is a zero dimensional Lie group.

A discrete group D is said to act properly discontinuously on a manifold M if the action $D \times M \to M$ satisfies two conditions

1) Each $x \in M$ has a neighbourhood U such that the set $\{a \in D \,|\, (aU) \cap U \neq \varnothing\}$ is finite;

2) If $x, y \in M$ are not in the same orbit of D, then there exist neighbourhoods U, V of x, y such that $U \cap DV = \varnothing$. Consider now an example. Let $M = S^2$ and $D = \mathbb{Z}_2 = \{e, a\}$. The group action $x \in S^2 \to e(x) = x$ and $a(x) = -x$ is free and properly discontinuous, and the quotient space $S^2/\mathbb{Z}_2$ is none other than $\mathbb{R}P(2)$, the real projective space of two dimensions.

The tiling of the plane $\mathbb{R}^2$ by square tiles, say, can be viewed as a discrete Lie group action on $\mathbb{R}^2$. The group D generated by reflections along the lines $x = 0$, $x = 1/2$; $y = 0$ and $y = 1/2$, relative to a fixed Cartesian system; it consists of discrete translations $(x, y) \mapsto (x + m, y + n)$, where m and n are integers.

We now state two important results.

Theorem:

All the isotropy groups H_x of points $x \in M$ are isomorphic if and only if M is a homogeneous space.   (E.36).

Theorem:

> There is a one-to-one correspondence between the points of a homogeneous space M (under the left action of G) and the left cosets G/H of H in G, where H is the isotropy group of G.   (E.37)

In connection with the topological classification of defects (Chap. 4) we defined the order parameter space as G/H. Theorem (E.37) is important in this context, and therefore we sketch the proof. Consider a point $x \in M$. Then with each left coset aH_x, we associate the point $f(a, x) \in M$. This correspondence can be shown to be the required isomorphism. When H is a closed subgroup of G (i.e., H is closed as a set), a natural topology called a quotient topology is induced by the topology of G on the set G/H so as to make it a manifold.

Some examples illustrating the use of theorem (E.37) are in order. The group O(3) acts on the homogeneous space S^2. The transitive nature of the group action is intuitively obvious. The isotropy group of the north pole $x = (0, 0, 1) \in S^2$ [imagine the Cartesian coordinate system (x^1, x^2, x^3) in $\mathbb{R}^3$ with S^2 given by the equation $(x^1)^2 + (x^2)^2 + (x^3)^2 = 1$] consists of matrices of the form

$$\left(\begin{array}{c|c} \underline{A} & \mathbf{0} \\ \hline \mathbf{0}^T & 1 \end{array} \right)$$

where $\underline{A} \in O(2)$, i.e., an orthogonal 2×2 matrix, and $\mathbf{0}^T = (0, 0)$. Hence, by using theorem (E.37), the space S^2 can be identified with O(3)/O(2). Since the group SO(3) is also transitive on S^2, analogous reasoning leads to the identification of S^2 with SO(3)/SO(2). This result is relevant in the physical example of three dimensional spin systems. Another example is that of the real projective space $\mathbb{R}P(2)$, which consists of points on S^2 with antipodal points identified. This example is pertinent to the study of defects in nematic liquid crystals. The action of O(3) [or SO(3)] on $\mathbb{R}P(2)$ is clearly transitive. The isotropy group H of the north pole $(0, 0, 1)$ is represented by matrices of the form

$$\left(\begin{array}{c|c} \underline{A} & \mathbf{0} \\ \hline \mathbf{0}^T & \pm 1 \end{array} \right) ,$$

where $\underline{A} \in O(2)$ [or SO(2)]. Clearly, $H = O(1) \otimes O(2)$ [or, $O(1) \otimes SO(2)$]. Hence, using theorem (E.36), we have the identification of $\mathbb{R}P^2$ and the space $O(3)/O(1) \otimes O(2)$ [or, the space $SO(3)/O(1) \otimes SO(2)$]. Note here that instead of the group O(1), we may take the group $\mathbb{Z}_2$ (see Sect. D.1) since it is isomorphic to O(1).

What we have discussed above will enable the reader to follow developments in this field. But the reader who actually wants to apply these ideas to problems of his or her interest, should necessarily know several other aspects of Lie groups and homotopy which have not been covered here. A few such aspects are: relative homotopy groups [E.1, 2, 4, 7] which are needed for the classification of surface defects, the concept of homology which has started finding application in defect physics [E.8], etc.

F. Local Gauge Invariance and Gauge Theories

Local gauge invariance plays a central role in modern physics. It is intimately related to the concept of a covariant derivative in a generalized sense (and thus to the "principle of minimal coupling" for interacting fields). There are numerous excellent articles and books on all aspects of gauge field theory (for a small sampling, see [F.1-6]. We have used, in this text, ideas related to local gauge symmetry and covariant derivatives in at least two contexts: those concerning the tessellation of curved spaces, defect lines on mapping to flat space, etc; and those involving "internal" connections (gauge fields) in the gradient terms of the Landau theory for superconductors, liquid crystals, etc. It is therefore helpful to recapitulate some of the basic ideas underlying gauge symmetry.

It has become increasingly clear that the natural mathematical language for this subject is that of differential geometry and fibre bundles. However, here we shall continue to use the more archaic language of vectors and tensors. We start with the concept of an internal connection which is extremely important for gauge theories.

F.1 Internal Connection

The connection introduced in Sect. B.2 may be referred to as an *external* connection since it is derived from a metric g_{ik}, which, in physical theories (like the theory of general relativity), is a function of space-time coordinates. There exists another type called the *internal connection*, which has nothing to do with the flatness or otherwise of a space-time manifold, but is connected with the internal degrees of freedom. A good example of this is isotopic spin, which enables us to regard the proton and the neutron as the 'up' and 'down' states of the same particle—the nucleon. The up or down is clearly not with respect to the laboratory, but an abstract (internal) space called the isotopic spin space. If the internal space is curved, then we need an internal connection.

Consider an n-component field $\psi_a(x)$ $(a = 1, \ldots, n)$ which is a scalar as far as space–time coordinates are concerned, but is a vector in the associated internal space of n dimensions. Suppose that under an infinitesimal transformation in the internal space, $\psi_a(x)$ transform as

$$\psi_a(x) \rightarrow \psi_a'(x) = \psi_a(x) + \theta^A(x) L^b_{Aa} \psi_b(x) \ . \tag{F.1}$$

Here $\{L^b_{Aa}\}$ are the generators of an internal (Lie) group (see Appendix D) and

$\{\theta^A\}$ are (infinitesimal) group parameters ($A = 1, \ldots, r$, the number of generators) which *may* depend on x. Let

$$\psi_a(x + dx) = \psi_a(x) + d\psi_a(x) = \psi_a(x) + \frac{\partial \psi_a}{\partial x^\mu} dx^\mu \ . \tag{F.2}$$

(Here the index μ runs over the space–time coordinates.) The rule for the parallel transport of the field ψ_a from x to $x + dx$ can be written as

$$\psi_a^{PT}(x + dx) = \psi_a(x) + \delta\psi_a(x) = \psi_a(x) + q\Gamma_{\mu a}^b \psi_b dx^\mu \tag{F.3}$$

where q is some constant, $\{\Gamma_{\mu a}^b\}$ are the coefficients of the relevant internal connection and the superscript PT denotes parallel transport. Note that, unlike the external connection of (B.13), $\Gamma_{\mu a}^b$ has both the space–time (μ) and the internal (a, b) indices. As we shall see in the next section, in gauge theories

$$\Gamma_{\mu a}^b = L_{Aa}^b B_\mu^A(x) \ , \tag{F.4}$$

where $\{B_\mu^A\}$ are the *gauge fields*. The "internal" covariant derivative is defined as

$$D_\mu \psi_a \equiv \frac{D\psi_a}{Dx^\mu} = \lim_{dx^\mu \to 0} \frac{\psi_a(x + dx) - \psi_a^{PT}(x + dx)}{dx^\mu} \ ,$$

which, on substitution of (F.2–4), leads to the expression:

$$D_\mu \psi_a = \partial_\mu \psi_a - q\Gamma_{\mu a}^b \psi_b \tag{F.5a}$$

$$= \partial_\mu \psi_a - qL_{Aa}^b B_\mu^A(x)\psi_b \tag{F.5b}$$

where $\partial_\mu \psi_a \equiv \partial\psi_a/\partial x^\mu$.

F.2 Gauge Field Theory

The framework in which the formalism is usually developed is that of Lagrangian field theory. One starts with a scalar (Lorentz-invariant) Lagrangian density $\mathscr{L}_0[\psi, \partial_\mu \psi]$ for an n-component "matter" field ψ_a (abbreviated by a vector ψ); the index μ runs over the space–time dimensions d (e.g., $\mu = 0, 1, 2, 3$ in the usual case). A variational principle (that of extremal action) yields the dynamical equation (the Euler–Lagrange equation)

$$\frac{\delta \mathscr{L}_0}{\delta \psi} = \partial_\mu \left[\frac{\delta \mathscr{L}_0}{\delta(\partial_\mu \psi)} \right] \ . \tag{F.6}$$

A symmetry transformation of the system is a transformation of ψ which leaves the action, and thence (F.6), invariant. ($\mathscr{L}$ itself may alter by a term that is a pure divergence). The *set* of solutions of (F.6) is thus invariant under a group of symmetry transformations. For a continuous symmetry group, this implies, under the conditions of Noether's theorem [F.7], a divergence-free current, a conserved quantity corresponding to it, etc.

In the condensed-matter context with which we have been concerned, the basic quantity in Landau-type theories is the free energy density $\mathscr{F}_0[\psi, \partial_\mu \psi, \dots]$. Here ψ is the (n-component) order parameter; $\mu = 1$, $2, \dots, d$ where d is the spatial dimensionality of the system. The gradient terms represent the effects of fluctuations. In principle, higher-order gradients may also occur in $\mathscr{F}_0$, but it is customary to restrict the expression to quadratic terms in the gradients. The interest here is in stable equilibrium states, which are given by the minima of $\mathscr{F}_0$, i.e., by the extremum conditions

$$\delta \mathscr{F}_0 = 0 , \quad \delta^2 \mathscr{F}_0 > 0 \tag{F.7}$$

(F.7) is the counterpart of (F.6). For example, in the expansion

$$\mathscr{F}_0 = a(T - T_c)\psi^2 + c\psi^4 + \tfrac{1}{2}K|\nabla\psi|^2 \quad (a, c, K > 0) \tag{F.8}$$

(Chap. 6) involving a scalar order parameter ψ, $\mathscr{F}_0$ is a functional of ψ and $\nabla\psi$. It is easy to verify that $\nabla\psi = 0$, $\psi = 0$ is the state of stable equilibrium for $T > T_c$; for $T < T_c$, the states $\nabla\psi = 0$, $\psi = \pm [a(T_c - T)/2c]^{1/2}$ are the states of stable equilibrium, as noted in the text.

The remarks following (F.6) on symmetry transformations apply to $\mathscr{F}_0$ and to the set of solutions of (F.7) as well. In what follows, this correspondence between $\mathscr{L}_0$ and $\mathscr{F}_0$, and between the field and the order parameter (both denoted by ψ), is to be understood.

F.3 U(1) Gauge Symmetry

This is the simplest example involving a continuous symmetry group. We consider a complex scalar field (order parameter) ψ with

$$\mathscr{L}_0 (\text{or } \mathscr{F}_0) = \Phi[|\psi|^2] + (\partial^\mu \psi)^* (\partial_\mu \psi) . \tag{F.9}$$

Here the raising operation $\partial^\mu = g^{\mu\nu}\partial_\nu$ is performed with the Minkowski metric $g^{\mu\nu}(g^{11} = g^{22} = g^{33} = -1, g^{00} = 1)$. Let us now consider the one-parameter group of (global) gauge transformations (of the field) given by

$$\psi \to g(\theta)\psi = \exp(iq\theta)\psi , \quad \psi^* \to \exp(-iq\theta)\psi^* , \tag{F.10}$$

where θ is a real number [evidently, $g \in U(1)$]. As the transformation is directly on the field, and not induced by a coordinate transformation, we note that $\partial_\mu \psi \to \partial_\mu(g\psi)$ under the transformation. It is then easy to see that $\mathscr{L}_0$ is invariant under such a transformation. U(1) is the symmetry group of the system.

We shall now examine whether the system can be made invariant under a *local* gauge transformation

$$\psi \to g\psi = \exp[iq\theta(x)]\psi , \quad \psi^* \to \exp[-iq\theta(x)]\psi^* , \tag{F.11}$$

where $\theta(x)$ is a real-valued function of the coordinates. The group of transformations is still U(1), but the element of U(1) that acts on ψ may differ from point to point. As $|\psi|$ is unaffected by a transformation of this sort, it is only the gradient term in (F.9) that we must examine. We have, as before, $\partial_\mu \psi \to \partial_\mu (g\psi)$. Owing to the x-dependence of g, however, $\partial_\mu (g\psi) = g\partial_\mu \psi + iq(\partial_\mu \theta)g\psi$. The presence of the second term prevents $\partial_\mu \psi$ from transforming precisely like ψ itself, and vitiates the invariance of $\mathscr{L}_0$ under this "local" group of transformations. The gradient term probes the behaviour of $\psi(x)$ in a *neighbourhood of x*. The fact that the field transforms in a different manner at different points shows up in the occurrence of the extra term in the transform of $\partial_\mu \psi$. If the derivative could be modified so as to transform in the same manner as the field itself, local gauge invariance would obtain. It turns out that the right choice of the derivative is an *internal* covariant derivative (F.5)

$$D_\mu \psi = (\partial_\mu - iq A_\mu)\psi \ . \tag{F.12}$$

The replacement rule $\partial_\mu \to D_\mu$ given by (F.12) is known as the principle of *minimal coupling*. Comparing with (F.5), we observe that the sole generator L of the one parameter group U(1) is the complex number i and A_μ is the gauge field—here known as the four-vector potential. A simple inspection shows that under the transformation (F.11),

$$D_\mu \psi \to \exp[iq\theta(x)]D_\mu \psi \quad \text{and}$$

$$(D_\mu \psi)^* \to \exp[-iq\theta(x)](D_\mu \psi)^* \ .$$

The minimally coupled Lagrangian is $\tilde{\mathscr{L}}_0 = \mathscr{L}_0(\psi, D_\mu \psi)$. It is easily verified that $\tilde{\mathscr{L}}_0$ is invariant under the transformation (F.11) if and only if $D_i \psi$ transforms according to $D_\mu \psi \to g(D_\mu \psi)$, which reduces to the requirement that the vector field $A_\mu(x)$ simultaneously transform according to

$$A_\mu \to A_\mu + \partial_\mu \theta \ . \tag{F.13}$$

Equations (F.11) and (F.13) define a local gauge transformation under which the Lagrangian (or free energy) density $\tilde{\mathscr{L}}_0$ is invariant, provided the gradient terms in it involve the covariant derivative as defined in (F.12).

In cases where the dynamics of the gauge field itself is also relevant, one must add the "pure gauge field" contribution to the Lagrangian which is invariant under the transformation (F.13). Noting that (F.13) involves the addition of a gradient, this contribution must be a functional of the curl:

$$F_{\mu\nu} = \partial_\mu A_\nu - \partial_\nu A_\mu \ . \tag{F.14}$$

Up to terms quadratic in A_μ, this contribution must therefore be proportional to $F_{\mu\nu}F^{\mu\nu}$. Then the total Lagrangian $\mathscr{L}$ can be written

$$\mathscr{L} = \tilde{\mathscr{L}}_0 - \tfrac{1}{4}F_{\mu\nu}F^{\mu\nu} \ . \tag{F.15}$$

Note that the commutator of the covariant derivatives defined in (F.12) is

$$[D_\mu, D_\nu]\psi_a = -\,\mathrm{i}\,F_{\mu\nu}\psi_a \;, \tag{F.16}$$

in accordance with the remarks made following (B.17).

The Euler–Lagrange equations derived from (F.15) lead to the coupled field equations

$$D_\mu D^\mu \psi = -\,\psi\,V'(\psi^*\psi) \;, \qquad V'(x) = \frac{\partial V}{\partial x} \;, \tag{F.17}$$

$$\partial_\mu \partial^\mu A^\nu - \partial^\nu \partial_\mu A^\mu = q j^\nu \tag{F.18a}$$

$$j^\nu = \mathrm{i}[\psi^*(D^\nu\psi) - (D^\nu\psi)^*\psi] \;. \tag{F.18b}$$

Observe that (F.18a) is identical to the Maxwell equations of electrodynamics (in the Lorentz gauge). Thus the principle of local U(1) gauge invariance leads to the "discovery" of the fundamental equations of the electromagnetic field.

We close this section with the observation that covariant derivatives have been used in the text in the case of "external" connections arising from spatial curvature, as in Sects. (5.6) and (6.5), as well as "internal" connections, as in the Landau–Ginzburg theory for type II superconductors (Sect. 6.6.5), liquid crystals (Sect. 6.6.3), etc.

F.4 Non-Abelian Gauge Groups

The idea of local gauge invariance can be extended to the case of a Lie group G of transformations acting on an n-dimensional multiplet of fields ψ. A gauge transformation on ψ is given by

$$\psi \to g(\theta)\psi = \exp(L_A \theta^A)\psi \;, \tag{F.19}$$

where $\theta = \{\theta^A\}$, $A = 1, 2, \ldots, r$ is a set of r real parameters, r being the dimension of the Lie algebra; $\{L_A\}$ are $(n \times n)$ skew-Hermitian matrix representations of the infinitesimal generators of G, obeying the commutation relations

$$[L_A, L_B] = C_{AB}^E L_E \;, \tag{F.20}$$

where $\{C_{AB}^E\}$ are the structure constants of the Lie algebra of G and the capital Roman letters range from 1 to r. The Lagrangian

$$\mathscr{L} = \Phi[\psi^\dagger \psi] + (\partial_\mu \psi)^\dagger (\partial^\mu \psi) \tag{F.21}$$

(where $\dagger$ denotes Hermitian conjugation) is invariant under the transformation (F.19), provided the parameters $\{\theta^A\}$ are constants. Local gauge transformations correspond to the case $\theta^A = \theta^A(x)$. To achieve invariance under such transformations, we must introduce a set of gauge fields (one for each dimension of the gauge group) $B_\mu = \{B_\mu^A\}$, and define an (internal) covariant derivative

$$D_\mu \psi = (\mathbb{1} \partial_\mu - L_A B_\mu^A) \psi \ , \tag{F.22}$$

where $\mathbb{1}$ is the $r \times r$ identity matrix. The Lagrangian is modified to

$$\tilde{\mathscr{L}}_0 = \Phi[\psi^\dagger \psi] + (D_\mu \psi)^\dagger (D^\mu \psi) \ . \tag{F.23}$$

This is invariant under the local gauge transformation (F.19) provided $(D_\mu \psi) \to g(\theta)(D_\mu \psi)$. In turn, this requires the following transformation rule for the gauge fields $\{B_\mu^A\}$:

$$(L_A B_\mu^A) \to g(\theta)[L_A B_\mu^A + g^{-1}(\theta)\partial_\mu g(\theta)]g^{-1}(\theta) \ . \tag{F.24}$$

It is easy to check that this reduces to (F.13) in the case of $G = U(1)$, $\psi = \psi$ (i.e., $n = 1$). The form (F.24) may give the impression that the transformation of B_μ^A is representation-dependent, because of the appearance of the matrices $\{L_A\}$ in it. However, if we consider an infinitesimal transformation $\psi \to \psi + L_A \theta^A \psi$, it can be shown that

$$B_\mu^A \to B_\mu^A + \partial_\mu \theta^A + C_{BD}^A \theta^B B_\mu^D \ . \tag{F.25}$$

Since the structure constants $\{C_{BD}^A\}$ are representation-independent, so is (F.25).

The final term on the right in (F.25) arises from the noncommutativity of the generators of the gauge group.

The commutator of covariant derivatives turns out to be

$$[D_\mu, D_\nu] \psi = L_A F_{\mu\nu}^A \psi \tag{F.26}$$

where the field strength is given by

$$F_{\mu\nu}^A = \partial_\mu B_\nu^A - \partial_\nu B_\mu^A - C_{DE}^A B_\mu^D B_\nu^E \ . \tag{F.27}$$

In contrast to the Abelian case, the field strength is no longer invariant under a gauge transformation; rather, it is "gauge covariant". Under an infinitesimal transformation,

$$F_{\mu\nu}^A \to F_{\mu\nu}^A + C_{DE}^A \theta^D F_{\mu\nu}^E \ . \tag{F.28}$$

What must be noted is the fact that this transformation law is a *homogeneous* one, in contrast to that for the gauge field itself. The latter has an inhomogeneous part [the final terms in (F.13) and (F.24), or the second term on the right in (F.25)]. This property has a deeper significance that is best expressed in terms of the curved space analogues. The connection Γ_{lm}^k (the analogue of the gauge field B_μ^A) is not a tensor in the usual sense of the word (i.e., under homogeneous transformations such as rotations); but the curvature tensor R_{jkl}^i (the analogue of $F_{\mu\nu}^A$) *is* such a tensor. (The connection is a tensor only if we restrict ourselves to transformations for which the inhomogeneous term in the transformed value of Γ^k_{lm} vanishes, i.e., to *affine* transformations.) In the present gauge field context, the difference may be stated also as follows: $F^A_{\mu\nu}$ transforms like a representation of the gauge group; B^A_μ, on the other hand, takes values from the Lie algebra of G. (See Appendix E.)

F.5 Gauge Theory of Dislocations and Disclinations

Recently there have been several attempts [F.8–15] to develop gauge theories of topological defects in condensed matter. As an example, we consider here a gauge theory of dislocations and disclinations in an elastic continuum. *Kadic* and *Edelen* [F.9] have proposed such a gauge theory in the spirit of the discussion outlined in Sects. F.1–4. Consider the Lagrangian of a perfect elastic continuum

$$\mathscr{L}_0 = \mathscr{L}_0(\{\partial_j u_i\}) = -\tfrac{1}{2}[\lambda(e_{ii})^2 + 2\mu e_{ik}e_{ik}] \ , \tag{F.29}$$

where $u_i(x)$ is the displacement field of the elastic continuum whose coordinates are denoted by x_i $(i = 1, 2, 3)$, e_{ik} is the strain tensor

$$e_{ik} = \tfrac{1}{2}(u_{i,k} + u_{k,i}) \ , \tag{F.30}$$

and (λ, μ) are the Lamé constants. The Lagrangian $\mathscr{L}_0$ is invariant under the global gauge group of transformations

$$u \rightarrow u' = g(\theta, b)u = (\underline{R}_\theta, \underline{T}_b)u$$

$$= \underline{R}(\theta)u + b \tag{F.31}$$

where $\underline{R}^T\underline{R} = \underline{R}\underline{R}^T = \underline{I}$ and $\partial_i\underline{R} = Q = \partial_i b$. Here $\{\underline{R}_\theta\}$ denotes the set of (orthogonal) matrices depending on the three rotation angles $\{\theta_k\}$ about the three Cartesian coordinate axes and $b = (b_1, b_2, b_3)^T$ represents the parameters of the translation group T(3). Thus the gauge group is the Euclidean group E(3) = T(3) $\wedge$ SO(3), where $\wedge$ denotes the semi-direct product. *Kadic* and *Edelen* [F.9] use E(3) as an internal symmetry group, i.e., the coordinates x_i remain unchanged under the group transformations whereas the fields $u_i(x)$ transform as $u_i \rightarrow u'_i = R_{ik}u_k + b_i$. When R_{ik} and b_i are made functions of x, i.e., when the transformations are made local, the Lagrangian $\mathscr{L}_0$ no longer remains invariant owing to the appearance of additional terms involving $\partial_i\theta_k(x)$ and $\partial_i b_k(x)$. As discussed earlier, the gauge invariance can be restored by introducing compensating (gauge) fields—in this case, six of them, via appropriate internal covariant derivatives involving these fields and the generators of SO(3) and T(3). Proceeding in this manner *Kadic* and *Edelen* [F.9] obtain the solutions of the field equations and claim that the gauge fields associated with the group T(3) describe dislocations whereas those with the group SO(3) describe the disclinations. Some of their results can be paraphrased as follows:

1) The stress field of a dislocation reduces to the classically known result only locally around the defect, but falls to zero rapidly for large distances,
2) The Burgers vector, calculated by taking the line integral of the displacement field around a closed loop surrounding the dislocation, tends to zero as the loop size increases,
3) The Peach–Koehler force between two dislocations decreases exponentially for large separations between them.

These results are unphysical, and although Kadic and Edelen offer some explanations, *Nabarro* [F.16] and *Kunin* [F.10] have raised serious objections.

The Kadic and Edelen approach has recently been carefully examined [F.17] and shortcomings in it have been eliminated. It has further been found that *both* dislocations and disclinations arise due to the breaking of the global T(3) invariance alone. We present below a short summary of this recent work.

The local gauge transformations corresponding to the action of T(3) are

$$u_k \to u'_k = u_k + b_k(x) \ , \qquad b_{k,l} \neq 0 \ . \tag{F.32}$$

The covariant derivative appropriate for the group is [F.17]

$$D_i u_k = \partial_i u_k + \varphi_{ik} \ . \tag{F.33}$$

Observe that the gauge fields φ_{ik} appear in an inhomogeneous manner as a consequence of the inhomogeneous group action (F.32), i.e., the second term in (F.33) does not contain $\{u_k\}$ unlike the construction (F.5). The fields φ_{ik} transform simultaneously with (F.32) according to

$$\varphi_{ik} \to \varphi'_{ik} = \varphi_{ik} - \partial_i b_k \tag{F.34}$$

in order to maintain the local gauge invariance of $\tilde{\mathscr{L}}_0(D_i u_k)$. This minimally coupled matter field Lagrangian is given by

$$\tilde{\mathscr{L}}_0 = -\tfrac{1}{2}[\lambda(E_{ii})^2 + 2\mu E_{ik} E_{ik}] \ , \tag{F.35a}$$

where

$$E_{ik} = e_{ik} + \tfrac{1}{2}(\varphi_{ik} + \varphi_{ki}) \ . \tag{F.35b}$$

The field tensor F_{aik} [analogous to $F_{\mu\nu}$ of (F.14)] is given by the curl of the potential,

$$F_{aik} = \varphi_{ak,i} - \varphi_{ai,k} \ . \tag{F.36}$$

The total Lagrangian is now

$$\mathscr{L} = \tilde{\mathscr{L}}_0(D_i u_k) - \frac{s}{2} F_{aik} F_{aik} \ , \tag{F.37}$$

where s is a coupling parameter. The field equations which follow from (F.37) are:

$$u_{i,kk} + (L+1)u_{k,ki} = -(\varphi_{ki,k} + \varphi_{ik,k} + L\varphi_{kk,i}) \tag{F.38a}$$

$$\varphi_{ik,ll} - \varphi_{il,lk} - \kappa^2(\varphi_{ik} + \varphi_{ki} + L\varphi_{ll}\delta_{ik})$$
$$= \kappa^2(u_{i,k} + u_{k,i} + Lu_{l,l}\delta_{ik}) \tag{F.38b}$$

where $L = \lambda/\mu$ and $\kappa^2 = \mu/2s$. The stress tensor σ_{ik} is obtained from $\mathscr{L}$ as

$$\sigma_{ik} = -\frac{\partial \mathscr{L}}{\partial u_{i,k}} = \lambda(u_{l,l} + \varphi_{ll})\delta_{ik} + \mu(u_{i,k} + u_{k,i} + \varphi_{ik} + \varphi_{ki}) \ . \tag{F.39}$$

It is necessary now to relate the φ-field to the dislocation density tensor. Recall first the definition of the Burgers vector (4.6) in the conventional elastic continuum theory:

$$b_k = \oint \partial_i u_k dx^i \equiv \oint \beta_{ki} dx^i \ . \tag{F.40}$$

Unlike the u-field occurring in the Lagrangian (F.37), the u-field above is *not* a single-valued continuous function of x. The quantity $\beta_{ki} = \partial_i u_k$ is the distortion tensor. Another way of expressing (F.40) is to convert the line integral on the rhs to a surface integral by using Stokes' theorem, obtaining

$$b_k = \oiint \alpha_{ik} dS_i \ . \tag{F.41a}$$

Here dS is an area element of the closed surface whose boundary coincides with the loop occurring in the integral of (F.40) and α_{ik}, the dislocation density tensor [F.18], is the curl of the tensor $\underset{\sim}{\beta}$:

$$\alpha_{ik} = \varepsilon_{iml}\beta_{kl,m} = \varepsilon_{iml}u_{k,lm} \ . \tag{F.42b}$$

[ε_{iml} is the Levi–Civita tensor already defined immediately after (6.60)]. The vanishing of the $\underset{\sim}{\alpha}$-tensor is precisely the condition for the existence of a continuous single-valued u-field. This suggests separating the contribution of the gradient $u_{i,k}$ in (F.40) into a completely integrable part $\bar{u}_{i,k}$ and a non-integrable part, φ_{ik}:

$$u_{i,k} = \bar{u}_{i,k} + \varphi_{ik} \ . \tag{F.43}$$

Referring back to the minimal coupling prescription (F.33), we realize that this is exactly what has been done in the gauge theory, provided we identify $\bar{u}_i$ with u_i in (F.33) and φ_{ik} with the gauge field. Thus the gauge field carries complete information about the defects (i.e., the non-integrable part of β_{ik}). With this understanding, the substitution of (F.33) in (F.42b) leads to the relation

$$\alpha_{ik} = \varepsilon_{iml}\varphi_{kl,m} \ , \tag{F.44}$$

which is the analogue of the familiar relation $B = V \times A$ of electromagnetic theory.

For a single screw dislocation lying along the x^3-axis, the Burgers vector is completely in the x^3-direction and the cylindrical symmetry of the situation requires the following conditions on u_i and α_{ik}:

$$u_1 = u_2 = 0 , \quad u_3 = u_3(\varrho) ,$$

$$\alpha_{ik} = \delta_{i3} \delta_{k3} \alpha(\varrho) , \tag{F.45}$$

where ϱ denotes the radial variable $\varrho = \sqrt{x_1^2 + x_2^2}$ and $\alpha(\varrho)$ is a scalar function of ϱ alone. The precise relation of α to the φ-field is

$$\alpha = \varphi_{32,1} - \varphi_{31,2} , \tag{F.46}$$

and all components of φ_{ik} other than those occurring in (F.46) can be chosen to be zero (by a suitable choice of the gauge). The field equations (F.38) then reduce to the form

$$u_{3,11} + u_{3,22} = -\varphi_{31,1} - \varphi_{32,2} , \tag{F.47a}$$

$$\varphi_{31,22} - \varphi_{32,21} - \kappa^2 \varphi_{31} = \kappa^2 u_{3,1} , \tag{F.47b}$$

$$\varphi_{32,11} - \varphi_{31,12} - \kappa^2 \varphi_{32} = \kappa^2 u_{3,2} . \tag{F.47c}$$

Differentiating (F.47c) by x_1, (F.47b) by x_2, and taking the difference, one obtains the equation

$$\alpha_{,11} + \alpha_{,22} = \kappa^2 \alpha , \tag{F.48}$$

the cylindrically symmetric solution of which is

$$\alpha(\varrho) = c\, K_0(\kappa\varrho) , \tag{F.49}$$

where C is a constant and K_0 is the modified Bessel function of the second kind and of order zero.

The Burgers vectors $b(R)$, obtained by using (F.41a), is

$$b(R) = 2\pi \int \alpha(\varrho) \varrho \, d\varrho$$

$$= \frac{2\pi C}{\kappa^2} [1 - \kappa R\, K_1(\kappa R)] , \tag{F.50}$$

where use has been made of the identity $\int K_0(x) x\, dx = -x K_1(x)$. Noting that $K_0(x) \sim -\ln x$, $K_1(x) \sim x^{-1}$ for $x \ll 1$ and $K_n(x) \sim \exp(-x)/\sqrt{x}$ $(n = 0, 1)$ for $x \gg 1$, it is seen from (F.50) that $b(0) = 0$ and $b(\infty) = 2\pi C/\kappa^2$ (thus determining the constant C). It is clearly natural to interpret κ^{-1} as the radius of the dislocation core which must necessarily be finite. This natural emergence of the core of the dislocation (as opposed to the ad hoc manner in which it is introduced in the conventional continuum approach [F.18]) is a pleasing feature of this theory [F.17].

Turning now to the stress field of the screw dislocation, this is found from (F.39) to be given by:

$$\sigma_{31} = \sigma_{13} = -\frac{\mu b(\infty)}{2\pi\varrho} \sin\theta [1 - \kappa\varrho K_1(\kappa\varrho)] \ ,$$

$$\sigma_{32} = \sigma_{23} = \frac{\mu b(\infty)}{2\pi\varrho} \cos\theta [1 - \kappa\varrho K_1(\kappa\varrho)] \ . \tag{F.51}$$

All other components of σ_{ik} vanish. It may be noted that the results of [F.9] differ from (F.51) in that the significant term 1 inside the square parentheses is absent in the former case. Observe that the stress field is the classical stress field [F.19] modulated by the factor given in the square parentheses. For $\varrho \to \infty$, (F.51) reduces to the correct classical solution, thus removing the undesirable feature of the Kadic and Edelen theory [F.9]. Note also that the stresses vanish at $\varrho = 0$, thus removing the divergency of the classical elasticity result [F.19]. The self-energy W of the dislocation turns out to be

$$W \approx \frac{\mu b^2(\infty)}{4\pi} [1 - \kappa R K_1(\kappa R)]^2 \ln(\kappa R) \ , \qquad (\kappa R \gg 1) \ .$$

The logarithmic divergence for $R \to \infty$ is expected on physical grounds.

It is possible to show [F.17] that choosing the nonvanishing components of the φ-field as $\varphi_{11} = \varphi_{22} = \varphi(\varrho)$ (a scalar function) leads to the solution for u and $\{\sigma_{ik}\}$ corresponding to a wedge disclination. (For a good review of the classical continuum theory of disclinations, see *De Wit* [F.18].) The qualitative features of the removal of the stress divergency at the origin, the reduction of the stress field to the classical form at large distance, the existence of a core, etc. are similar to those in the case of a dislocation. Since disclinations are rotational analogues of dislocations, intuition may suggest SO(3) as the (internal) gauge group, which is what Kadic and Edelen assume. However, their solution for the gauge field has spherical symmetry and *not* the cylindrical symmetry which one *should* expect for a line defect like a disclination.

In the Kadic–Edelen approach, disclinations arise due to the local action of the gauge group SO(3). Now the homotopy group Π_1 [SO(3)] is the group $\mathbb{Z}_2$, the group of the integers modulo two. However, one knows that the Frank vectors of the disclinations add arithmetically like integers. This deficiency of the Kadic–Edelen approach is also removed in the present theory [F.17], since in this theory the disclinations (like the dislocations) arise in the context of the local action of the Abelian group T(3).

References

Chapter 2

2.1   P. M. de Wolff: Acta Crystallogr. A **30**, 777 (1974)
2.2   D. Shechtman, I. Blech, D. Gratias, J. W. Cahn: Phys. Rev. Lett. **53**, 1951 (1984)
2.3   M. Gardner: Sci. American **236/1**, 110 (1977)
2.4   D. Levine, P. J. Steinhardt: Phys. Rev. B **34**, 596 (1986)
2.5   J. E. S. Socolar, P. J. Steinhardt: Phys. Rev. B **34**, 617 (1986)
2.6   R. K. P. Zia, W. J. Dallas: J. Phys. A **18**, L341 (1985)
2.7   N. G. de Bruijn: Akad. Wetensch. Proc. Ser. A **43**, 39 (1981)
2.8   J. E. S. Socolar, P. J. Steinhardt, D. Levine: Phys. Rev. B **32**, 5547 (1985)
2.9   V. Kumar, D. Sahoo, G. Athithan: Phys. Rev. B **34**, 6924 (1986)
2.10 T. L. Ho: Phys. Rev. Lett. **56**, 468, 1927 (1986)
2.11 "Workshop on Aperiodic Crystals", Les Houches, March 11–20, 1986, published in J. de Phys. **47**, C3–9 (1986)
2.12 P. G. De Gennes: *The Physics of Liquid Crystals* (Oxford Univ. Press, London 1974)
2.13 S. Chandrasekhar: *Liquid Crystals* (Cambridge Univ. Press, Cambridge 1977)
2.14 S. Chandrasekhar, D. K. Sadashiva, K. A. Suresh: Pramana **9**, 471 (1977)
2.15 B. I. Halperin, D. R. Nelson: Phys. Rev. Lett. **41**, 121 (1978)
2.16 G. Venkataraman, D. Sahoo: Pramana **24**, 317 (1985)
2.17 R. Zallen: *Physics of Amorphous Solids* (Wiley, New York 1983)
2.18 J. D. Bernal: Proc. R. Soc. A **280**, 299 (1964)
2.19 J. D. Bernal, S. V. King: "Experimental Studies of a Simple Liquid Model", In *Physics of Simple Liquids*, ed. by H. N. V. Temperly, J. S. Rowlinson, G. S. Rushbrooke (North Holland, Amsterdam 1968) pp. 233–252
2.20 C. H. Bennett: J. Appl. Phys. **43**, 2727 (1972)
2.21 J. F. Sadoc, J. Dixmier, A. Guinier: J. Non-Cryst. Solids **12**, 46 (1973)
2.22 F. Ordway: Science **141**, 800 (1964)
2.23 D. L. Evans, S. V. King: Nature **212**, 1353 (1966)
2.24 D. E. Polk: J. Non-Cryst. Solids **5**, 365 (1971)
2.25 P. Chaudhari, J. F. Graczyk, D. Henderson, P. Steinhardt: Phil. Mag. **31**, 727 (1975)
2.26 B. I. Halperin: "Theory of Melting and Liquid Crystal Phases in Two Dimensions", In *Symmetry and Broken Symmetries*, ed by N. Boccara (Idset, Paris 1981) p. 183

Chapter 3

3.1   D. Ter Haar (ed.): *Collected papers of L. D. Landau* (Pergamon, London 1965) p. 193
3.2   L. D. Landau, E. M. Lifshitz: *Statistical Physics* (Pergamon, Oxford 1959)
3.3   See, for example, D. Forster: *Hydrodynamic Fluctuations, Broken Symmetry and Correlation Functions* (Benjamin, New York 1975)
3.4   G. E. Volovik, V. P. Mineev: Zh. Eksp. Teor. Fiz. **72**, 2256 (1976) [English transl.: Sov. Phys. JETP **45**, 1186 (1977)]

3.5   K. G. Wilson: Rev. Mod. Phys. **55**, 583 (1983)
3.6   P. W. Anderson: *Basic Notions of Condensed Matter Physics* (Benjamin/Cummins, Menlo Park, Calif. 1984)
3.7   H. Schmidt: "The Phases of Condensed Matter", In *Symmetries and Broken Symmetries*, ed. by N. Boccara (Idset, Paris 1981) p. 123
3.8   P. W. Anderson: *Concepts in Solids* (Benjamin, New York 1963)
3.9   J. Goldstone: Nuovo Cim. **19**, 154 (1961)
3.10 L. D. Landau: Phys. Z. Sowjet Union **11**, 26 (1937); see also E. M. Lifshitz, L. P. Pitaevskii: *Statistical Physics* Pt. I (Landau–Lifshitz Course of Theoretical Physics, Vol. 5) 3rd ed. (Pergamon, Oxford 1980) p. 432
3.11 R. E. Peierls: Ann. Inst. Henri Poincare **5**, 177 (1935)
3.12 L. Onsager: Phys. Rev. **65**, 117 (1944)
3.13 N. D. Mermin, H. Wagner: Phys. Rev. Lett. **17**, 1133 (1966)
3.14 J. M. Kosterlitz, D. J. Thouless: J. Phys. C **6**, 1181 (1973)
3.15 D. R. Nelson, B. I. Halperin: Phys. Rev. B **19**, 2457 (1979)
3.16 B. I. Halperin: "Theory of Melting in Two Dimensions", In *Symmetries and Broken Symmetries*, ed. by N. Boccara (Idset, Paris 1981) p. 183
3.17 See the review G. Venkataraman, D. Sahoo: Pramana **24**, 317 (1985)
3.18 C. A. Murray, D. H. Van Winkle: Phys. Rev. Lett. **58**, 1200 (1987)

Chapter 4

4.1   R. Shankar: J. de Physique **38**, 1405 (1977); see also D. Finkelstein: J. Math. Phys. **7**, 1218 (1966)
4.2   G. E. Volovik, V. P. Mineev: Zh. Eksp. Teor. Fiz. **72**, 2256 (1977) [English transl.: Sov. Phys. JETP **45**, 1186 (1977)]
4.3   V. P. Mineev, G. E. Volovik: Phys. Rev. B **18**, 3197 (1978)
4.4   N. D. Mermin: Rev. Mod. Phys. **51**, 591 (1979)
4.5   H. R. Trebin: Adv. Phys. **31**, 195 (1982)
4.6   G. Venkataraman: Current Science (India) **54**, 1 (1985)
4.7   G. Toulouse, M. Kleman: J. Physique Lett. **37**, L149 (1976)
4.8   B. Julia, G. Toulouse: J. Physique Lett. **40**, L395 (1979)

Chapter 5

5.1   G. Toulouse: Commun. Phys. **2**, 115 (1977)
5.2   Quoted by G. Toulouse: "The Frustration Model", In *Modern Trends in the Theory of Condensed Matter*, ed. by A. Pekalski, J. Przystawa (Springer, Berlin, Heidelberg 1980) p. 195
5.3   M. Kleman, J. F. Sadoc: J. Physique **40**, L569 (1979)
5.4   G. Venkataraman, D. Sahoo: Contemp. Phys. **26**, 579 (1985); **27**, 3 (1986)
5.5   A. Janner, T. Janssen: Phys. Rev. B **15**, 643 (1977)
5.6   P. M. de Wolff: Acta Crystallogr. A **30**, 777 (1974)
5.7   V. Elser: Phys. Rev. B **32**, 4892 (1985)
5.8   M. Duneau, A. Katz: Phys. Rev. Lett. **54**, 2688 (1985); A. Katz, M. Duneau: J. Physique **47**, 181 (1986)
5.9   P. Kalugin, A. Yu Kitaev, L. S. Levitov: Pis'ma Zh. Eksp. Teor. Fiz. **41**, 119 (1985) [English transl.: JETP Lett. **41**, 145 (1985)]; J. Physique Lett. **46**, L601 (1985)
5.10 R. K. P. Zia, W. J. Dallas: J. Phys. A **18**, L341 (1985)
5.11 H. S. M. Coxeter, G. J. Whitrow: Proc. Roy. Soc. A **201**, 417 (1950)
5.12 H. S. M. Coxeter: *Regular Polytopes* (Dover, New York 1973)
5.13 F. C. Frank, J. S. Kasper: Acta Cryst. **11**, 184 (1958); **12**, 483 (1959)

5.14 F. C. Frank: Proc. Roy. Soc. A **215**, 43 (1952)
5.15 J. F. Sadoc, R. Mosseri: Phil. Mag. **45**, 467 (1982)
5.16 J. F. Sadoc: J. Non-Cryst. Solids **44**, 1 (1981)
5.17 R. Zallen: *Physics of Amorphous Solids* (Wiley, New York 1983)
5.18 J. D. Bernal: Proc. R. Soc. A **280**, 299 (1964)
5.19 J. F. Sadoc: J. Physique **44**, L707 (1983)
5.20 J. F. Sadoc, R. Mosseri: J. Physique **46**, 1809 (1985)
5.21 D. R. Nelson: Phys. Rev. B **28**, 5515 (1983)
5.22 D. Polk: J. Non-Cryst. Solids **5**, 365 (1971)
5.23 G. A. N. Connell, R. J. Temkin: Phys. Rev. B **9**, 5323 (1974)
5.24 J. F. Sadoc, R. Mosseri: J. Non-Cryst. Solids **61/62**, 487 (1984)
5.25 J. Sadoc: J. Physique Colloque C **8**, 326 (1980)
5.26 N. P. Warner: Proc. R. Soc. A **383**, 359, 379 (1982)
5.27 D. Nelson, M. Widom: Nucl. Phys. B **240**, 113 (1984)
5.28 M. Kleman: J. Physique **43**, 1389 (1982); **44**, L295 (1983)
5.29 M. Kleman, P. Donnadieu: Philos. Mag. B **52**, 121 (1985)
5.30 M. Kleman: J. Physique **46**, L723 (1985)
5.31 S. Meiboom, J. P. Sethna, P. W. Anderson, W. F. Brinkman: Phys. Rev. Lett. **46**, 1216 (1981);
      S. Meiboom, M. Sammon, W. F. Brinkman: Phys. Rev. A **27**, 438 (1983)
5.32 N. G. de Bruijn: Ned. Akad. Weten. Proc. Ser. A **43**, 39, 59 (1981)
5.33 P. Kramer, R. Neri: Acta Crystallogr. A **40**, 580 (1984)
5.34 D. Shechtman, I. Blech, D. Gratias, J. W. Cahn: Phys. Rev. Lett. **53**, 1951 (1984)
5.35 J. E. S. Socolar, P. J. Steinhardt: Phys. Rev. B **34**, 617 (1986)
5.36 J. E. S. Socolar, P. J. Steinhardt, D. Levine: Phys. Rev. B **32**, 5547 (1985)
5.37 P. Bak: Phys. Rev. B **32**, 5764 (1985)
5.38 M. Kleman, Y. Gefen, A. Pavlovitch: Europhys. Lett. **1**, 61 (1986)
5.39 J. Bohsung, H. R. Trebin: Phys. Rev. Lett. **58**, 1204 (1987)
5.40 M. C. Valsakumar: Private communication
5.41 D. Ruelle: Physica **113A**, 619 (1982)
5.42 D. Ruelle: Physica **7D**, 40 (1983)

Chapter 6

6.1   L. D. Landau: *Collected Papers of L. D. Landau*, ed. by D. Ter Haar (Pergamon, London 1965)
      p. 193
6.2   E. M. Lifshitz, L. P. Pitaevskii: *Landau–Lifshitz Course of Theoretical Physics: Statistical
      Physics Part I* (Pergamon, Oxford 1980)
6.3   J. P. Elliott, P. G. Dawber: *Symmetry in Physics* (Macmillan Press, London 1979) Vol. 2, p. 326
6.4   M. V. Jaric: Phys. Rev. Lett. **55**, 607 (1985)
6.5   S. Alexander, J. P. McTague: Phys. Rev. Lett. **41**, 702 (1978)
6.6   P. Bak: Phys. Rev. Lett. **54**, 1517 (1985); Phys. Rev. B **32**, 5764 (1985)
6.7   N. D. Mermin, S. Troian: Phys. Rev. Lett. **54**, 1524 (1985)
6.8   P. A. Kalugin, A. Yu. Kitaev, L. S. Levitov: Pis'ma Zh. Eksp. Teor. Fiz. **41**, 119 (1985) [English
      transl.: JETP Lett. **41**, 145 (1985)]; J. Physique Lett. **46**, L601 (1985)
6.9   M. V. Jaric: Preprint 1986
6.10 P. J. Steinhardt, D. R. Nelson, M. Ronchetti: Phys. Rev. B **28**, 784 (1983)
6.11 A. D. J. Haymet: Phys. Rev. B **27**, 1725 (1983)
6.12 D. R. Nelson, J. Toner: Phys. Rev. B **24**, 363 (1981)
6.13 Y. Fujii, S. Hoshino, Y. Yamada, G. Shirane: Phys. Rev. B **9**, 4549 (1974)
6.14 D. R. Nelson, M. Widom: Nuclear Phys. B **240**, 113 (1984)
6.15 S. Sachdev, D. R. Nelson: Phys. Rev. Lett. **53**, 1947 (1984)

6.16 S. Sachdev, D. R. Nelson: Phys. Rev. B **32**, 1480 (1985)
6.17 G. Venkataraman, D. Sahoo: Contemp. Phys. **27**, 1 (1986)
6.18 J. Proust: Adv. Phys. **33**, 1 (1984)
6.19 G. Grinstein, T. C. Lubensky, J. Toner: Phys. Rev. B **33**, 3306 (1986)
6.20 P. De Gennes: *The Physics of Liquid Crystals* (Oxford University Press, London 1974)
6.21 G. E. Volovik, N. P. Mineev: Zh. Eksp. Teor. Fiz. **72**, 2256 [English transl.: JETP **45**, 1186 (1977)]
6.22 P. G. De Gennes: Solid State Commun. **10**, 753 (1972)
6.23 M. Kleman: *Points, Lines and Walls* (Wiley, New York 1983)
6.24 B. I. Halperin, T. C. Lubensky, S. K. Ma: Phys. Rev. Lett. **32**, 2922 (1974)
6.25 P. C. Martin, O. Parodi, P. S. Pershan: Phys. Rev. A **6**, 2401 (1972)
6.26 D. Forster: *Hydrodynamic Fluctuations, Broken Symmetry and Correlation Functions* (Benjamin, New York 1975)
6.27 D. Levine, T. C. Lubensky, S. Ostlund, S. Ramaswamy, P. J. Steinhardt, J. Toner: Phys. Rev. Lett. **54**, 1520 (1985)
6.28 T. C. Lubensky, S. Ramaswamy, J. Toner: Phys. Rev. B **32**, 7444 (1985)
6.29 M. Born, K. Huang: *Dynamical Theory of Crystal Lattices* (Clarendon Press, Oxford 1954)
6.30 G. Venkataraman, L. A. Feldkamp, V. C. Sahni: *Dynamics of Perfect Crystals* (MIT Press, Cambridge Mass. 1974)
6.31 K. G. Wilson: Physica, **73**, 119 (1974)
6.32 K. G. Wilson: Rev. Mod. Phys. **55**, 583 (1983)
6.33 P. Pfeuty, G. Toulouse: *Introduction to the Renormalization Group and to Critical Phenomena* (Wiley, New York 1975)
6.34 J. B. Sethna, D. C. Wright, N. D. Mermin: Phys. Rev. Lett. **51**, 467 (1983)
6.35 S. Meiboom, J. P. Sethna, P. W. Anderson, W. F. Brinkman: Phys. Rev. Lett. **46**, 1216 (1981)
6.36 J. M. Kosterlitz, D. J. Thouless: J. Phys. C **6**, 1181 (1973);
     J. V. Jose, L. P. Kadanoff, S. Kirkpatrick, D. R. Nelson: Phys. Rev. B **16**, 1217 (1977)
6.37 J. B. Kogut: Rev. Mod. Phys. **51**, 659 (1979)
6.38 S. A. Solla, E. K. Riedel: Phys. Rev. B **23**, 6008 (1981)
6.39 D. R. Nelson, B. I. Halperin: Phys. Rev. B **19**, 2457 (1979)
6.40 C. A. Murray, D. H. Van Winkle: Phys. Rev. Lett. **58**, 1200 (1977)
6.41 R. Kerner: Phil. Mag. B **47**, 151 (1983)
6.42 N. Rivier, D. M. Duffy: J. Physique **43**, 293 (1982)
6.43 M. G. Brereton, S. Shah: J. Phys. A **13**, 2751 (1980)
6.44 H. R. Brandt, K. Kawasaki: J. Phys. C **19**, 937 (1986)

Chapter 7

7.1 S. Aubry: Physica **7D**, 240 (1982)
7.2 S. Aubry, P. Y. Le Daeron: Physica **8D**, 381 (1983)
7.3 S. Aubry: J. Physique **44**, 147 (1983)
7.4 P. Reichert, R. Schilling: Phys. Rev. B**30**, 917 (1984)
7.5 P. Reichert, R. Schilling: Phys. Rev. B**32**, 5731 (1985)
7.6 A. J. Lichtenberg, M. A. Lieberman: *Regular and Stochastic Motion* (Springer, Berlin, Heidelberg 1983)
7.7 R. H. G. Helleman: "Self-Generated Chaotic Behaviour in Nonlinear Mechanics", In *Fundamental Problems in Statistical Mechanics*, ed. by E. G. D. Cohen (North-Holland, Amsterdam 1980) p. 165
7.8 M. H. Jensen, P. Bak: Phys. Rev. B **27**, 6853 (1983)
7.9 P. L. Taylor: Materials Science Forum **4**, 105 (1985)
7.10 S. Aubry, C. Godreche: J. Physique **47**, C3-187 (1986)

Chapter 8

8.1   R. G. Palmer: Adv. Phys. **31**, 669 (1982)
8.2   D. Forster: *Hydrodynamic Fluctuations, Broken Symmetry and Correlation Functions* (Benjamin, New York 1975)
8.3   E. Fradkin, B. A. Huberman, S. H. Shenker: Phys. Rev. B **18**, 4789 (1978)
8.4   P. W. Anderson: In *Ill-Condensed Matter*, ed. by R. Balian, R. Maynard, G. Toulouse (North-Holland, Amsterdam 1983) 2nd ed., p. 159
8.5   S. F. Edwards: In *Order in Strongly Fluctuating Systems*, ed. by T. D. Riste (Plenum, New York 1980) p. 1

Chapter 9

9.1   L. Pauling, R. Hayward: *Architecture of Molecules* (Freeman, San Francisco 1964)
9.2   H. S. M. Coxeter: *Introduction to Geometry* (Wiley, New York 1969)
9.3   N. J. A. Sloane: Sci. Am. **250/1**, 92 (1984)
9.4   G. Ya. Lyubarskii: *The Application of Group Theory in Physics* (Pergamon, New York 1960)
9.5   T. S. Lomont: *Applications of Finite Groups* (Academic, New York 1959)
9.6   L. Michel: Rev. Mod. Phys. **52**, 617 (1980)
9.7   D. Kastler, G. Loupias, M. Mebkhout, L. Michel: Commun. Math. Phys. **27**, 195 (1972)
9.8   L. Michel: In *Symmetries and Broken Symmetries*, ed. by N. Boccara (Idset, Paris 1981) p. 21
9.9   I. Dzyaloshinskii: In *Symmetries and Broken Symmetries*, ed. by N. Boccara (Idset, Paris 1981) p. 29
9.10 H. A. Kramers: Proc. Acad. Amsterdam **33**, 959 (1930)
9.11 E. P. Wigner: Goett. Nachr. Math. Phys. **31**, 546 (1932)
9.12 L. D. Landau: In *Collected Papers of L. D. Landau*, ed. by D. Ter Haar (Pergamon, London 1965) p. 193
9.13 F. H. Busse: In *Pattern Formation by Dynamical Systems*, ed. by H. Haken, Springer Ser. Syn., Vol. 5 (Springer, Berlin, Heidelberg 1979) p. 56
9.14 P. Berge: In *Chaos and Order in Nature*, Springer Ser. Syn., Vol. 11 (Springer, Berlin, Heidelberg 1981) p. 14
9.15 R. E. Rosensweig: In *Evolution of Order and Chaos*, ed. by H. Haken, Springer Ser. Syn., Vol. 17 (Springer, Berlin, Heidelberg 1982) p. 52
9.16 D. H. Sattinger: *Group Theoretic Methods in Bifurcation Theory*, Lecture Notes Math. **762** (Springer, Berlin, Heidelberg 1979)
9.17 N. Rivier, R. Occelli, J. Pantaloni, A. Lissouwski: J. Physique **45**, 49 (1984)
9.18 J. Friedel: In *Symmetries and Broken Symmetries*, ed. by N. Boccara (Idset, Paris 1981) p. 197

Appendix

A.1   P. C. Martin, O. Parodi, P. S. Pershan: Phys. Rev. A **6**, 2401 (1972)
A.2   H. Haken: *Synergetics: An Introduction*, 3rd ed., Springer Ser. Syn., Vol. 1 (Springer, Berlin, Heidelberg 1983)
A.3   G. Nicolis, I. Prigogine: *Self-Organization in Nonequilibrium Systems* (Wiley, New York 1977)
A.4   S. R. deGroot, P. Mazur: *Nonequilibrium Thermodynamics* (North-Holland, Amsterdam 1963)
A.5   H. J. Kreuzer: *Nonequilibrium Thermodynamics and its Statistical Foundations* (Clarendon, Oxford 1981)
A.6   I. E. Dzyaloshinskii, G. E. Volovik: Ann. Phys. **125**, 67 (1980)
A.7   D. Forster: *Hydrodynamic Fluctuations, Broken Symmetry, and Correlation Functions* (Benjamin, New York 1975)

198   References

A.8   L. D. Landau, I. M. Lifshitz: *Fluid Mechanics* (Pergamon, Oxford 1959)

A.9   H. Grabert: *Projection Operator Techniques in Nonequilibrium Statistical Mechanics* , Springer Tracts Mod. Phys., Vol. 15 (Springer, Berlin, Heidelberg 1982)

A.10   T. R. Kirkpatrick. E. G. D. Cohen, J. R. Dorfman: Phys. Rev. A **26**, 995 (1982);
T. R. Kirkpatrick, E. G. D. Cohen: J. Stat. Phys. **33**, 639 (1983);
C. Tremblay, A. M. S. Tremblay: Phys. Rev. A **25**, 1692 (1982);
H. Pleiner, H. Brand: J. Phys. Lett. **44**, L-23 (1983)

A.11   D. N. Zubarev: *Nonequilibrium Statistical Thermodynamics* (Consultants Bureau, New York 1974)

A.12   L. E. Malvern: *Introduction to the Mechanics of a Continuous Medium* (Prentice-Hall, New Jersey 1969)

A.13   F. J. Belinfante: Physica **6**, 887 (1939)

A.14   H. J. Pleiner: J. Phys. C-10, 4241 (1977)

A.15   J. Meixner, H. G. Reik: In *Handbuch der Physik* III/2 (Springer, Berlin 1959)

A.16   J. A. McLennan: Phys. Rev. A **10**, 1272 (1974)

A.17   H. N. W. Lekkerkerker, W. G. Laidlaw: Phys. Rev. A **9**, 431 (1974)

A.18   H. Brand, H. Pleiner: J. de Phys. **41**, 553 (1980)

A.19   A. Griffin: Rev. Mod. Phys. **40**, 167 (1968)

A.20   P. D. Fleming, C. Cohen: Phys. Rev. B **13**, 500 (1976)

A.21   G. E. Volovik. E. I. Kats: Sov. Phys. JETP **54**, 122 (1981)

A.22   D. D. Holm: Physica **17D**, 1 (1985)

A.23   D. D. Holm, B. A. Kupershmidt, Phys. Lett. **105A**, 225 (1984)

B.1   B. Spain: *Tensor Calculus* (Oliver and Boyd, London 1960)

B.2   J. J. Stoker: *Differential Geometry* (Wiley, New York 1969)

B.3   B. A. Dubrovin, A. T. Fomenko, S. P. Novikov: *Modern Geometry—Methods and Applications,* Part I (Springer, Berlin, Heidelberg 1984)

B.4   K. Kondo: Proc. 2nd Japan Nat. Congr. Applied Mechanics, Tokyo, 1952;
B. A. Bilby, R. Bullough, E. Smith: Proc. R. Soc. A **231**, 263 (1955);
E. Kroner: In *Physics of Defects*, ed. by R. Balian, M. Kleman, J. P. Poirier (North-Holland, Amsterdam 1981) p. 219

B.5   M. Kleman, J. F. Sadoc: J. de Phys. Lett. **40**, 569 (1979);
M. Kleman: J. de Phys. Lett. **43**, 1389 (1982)

B.6   J. P. Sethna: Phys. Rev. B **31**, 6278 (1985)

C.1   D. R. Nelson: Phys. Rev. B **28**, 5515 (1983)

C.2   J. F. Sadoc, R. Mosseri: J. de Phys. **46**, 1809 (1985)

C.3   H. S. M. Coxeter: *Introduction to Geometry* (Wiley, New York 1969)

D.1   P. Roman: *Some Modern Mathematics for Physicists and Other Outsiders*, Vol. 1 (Pergamon, New York 1975)

D.2   M. Hammemesh: *Group Theory* (Addison-Wesley, Reading, Mass. 1962)

D.3   R. Gilmore: *Lie Groups, Lie Algebras and Some of Their Applications* (Wiley, New York 1974)

E.1   N. D. Mermin: Rev. Mod. Phys. **51**, 591 (1971);
H.-R. Trebin: Adv. Phys. **31**, 195 (1982);
V. P. Mineev: In *Soviet Scientific Reviews, Sec. A: Physics Reviews*, Vol. 2, ed. by I. M. Khalatnikov (Harwood, London 1980) p. 173

E.2   L. Michel: Rev. Mod. Phys. **52**, 617 (1980)

E.3   D. H. Sattinger, O. L. Weaver: *Lie Groups and Algebras with Applications to Physics, Geometry, and Mechanics* (Springer, Berlin, Heidelberg 1986)

E.4   C. Nash, S. Sen: *Topology and Geometry for Physicists* (Academic, London 1983)

E.5   B. A. Dubrovin, A. T. Fomenko, S. P. Novikov: *Modern Geometry—Methods and Applications,* Pt. II (Springer, New York 1985)

E.6   R. Gilmore: *Lie Groups, Lie Algebras, and Some of Their Applications* (Wiley, New York 1974)

E.7   S. T. Hu: *Homotopy Theory* (Academic, New York 1959)

E.8   H.-R. Trebin: Phys. Rev. B **30**, 4338 (1984)
F.1   L. D. Landau, E. M. Lifshitz: *The Classical Theory of Fields* (Addison-Wesley, New York 1961)
F.2   C. N. Yang, R. Mills: Phys. Rev. **96**, 191 (1954)
F.3   E. S. Abers, B. W. Lee: Phys. Reports **9C**, 19 (1973);
      R. Utiyama: Phys. Rev. **101**, 1597 (1956)
F.4   K. Moriyasu: *An Elementary Primer for Gauge Theory* (World-Scientific, Singapore 1984)
F.5   B. A. Dubrovin, A. T. Fomenko, S. P. Novikov: *Modern Geometry—Methods and Applications*,
      Pt. I (Springer, Berlin, Heidelberg 1984)
F.6   L. O'Raifeartaigh: *Group Structure of Gauge Theories* (Cambridge Univ. Press, Cambridge
      1986)
F.7   C. Itzykson, J. Zuber: *Quantum Field Theory* (McGraw-Hill, New York 1980)
F.8   A. A. Golibiewska-Lasota: Int. J. Eng. Sci. **17**, 329 (1979);
      A. A. Golibiewska-Lasota, D. G. B. Edelen, Int. J. Eng. Sci. **17**, 335 (1975)
F.9   A. Kadic, D. G. B. Edelen: *Gauge Theory of Dislocations and Disclinations*, (Springer, Berlin,
      Heidelberg 1983)
F.10 I. A. Kunin: In *The Mechanics of Dislocations*, ed. by E. C. Aifantis, J. P. Hirth, (American
      Society of Metals, Metals Park, Ohio 1985) p. 69
F.11 D. M. Duffy, N. Rivier: J. Physique **43** C9-475 (1982)
F.12 K. Kawasaki, H. R. Brand: Ann. Phys. **160**, 420 (1985)
F.13 B. K. D. Gairola: In *Continuum Models of Discrete Systems* **4**, ed. by O. Brulin, R. K. T. Hsieh
      (North-Holland, Amsterdam 1981)
F.14 E. Kroner: In *The Mechanics of Dislocations*, ed. by E. C. Aifantis, J. P. Hirth (American Society
      of Metals, Metals Park, Ohio 1985) p. 57
F.15 G. Venkataraman, D. Sahoo: Contemp. Phys. **27**, 3 (1986)
F.16 F. R. N. Nabarro: Private communication
F.17 M. C. Valsakumar, D. Sahoo: *Bull. Mater. Sci.* **10**, 3 (1988)
F.18 R. DeWit: J. Res. Nat. Bur. Stand. (U.S.) **77A**, 49, 359, 607 (1973)
F.19 J. P. Hirth, J. Lothe: *Theory of Dislocations*, (Wiley, New York 1982)

Author Index

Subject Index